Brain and Human Body Modelling 2021

Sergey Makarov • Gregory Noetscher
Aapo Nummenmaa

Editors

Brain and Human Body Modelling 2021

Selected papers presented at 2021 BHBM
Conference at Athinoula A. Martinos Center
for Biomedical Imaging, Massachusetts
General Hospital

 Springer

Editors
Sergey Makarov
ECE Department
Worcester Polytechnic Institute
Worcester, MA, USA

Gregory Noetscher
Worcester Polytechnic Institute
Worcester, MA, USA

Aapo Nummenmaa
Harvard Medical School
Massachusetts General Hospital
Charlestown, MA, USA

ISBN 978-3-031-15453-9   ISBN 978-3-031-15451-5   (eBook)
https://doi.org/10.1007/978-3-031-15451-5

This Springer imprint is published by the registered company Springer Nature Switzerland AG
The registered company address is: Gewerbestrasse 11, 6330 Cham, Switzerland

Contents

Part I Low Frequency Electromagnetic Modeling and Experiment: Tumor Treating Fields

The Impact of Scalp's Temperature in the Predicted LMiPD
in the Tumor During TTFields Treatment for Glioblastoma
Multiforme. 3
Nichal Gentilal, Ariel Naveh, Tal Marciano, Zeev Bomzon,
Yevgeniy Telepinsky, Yoram Wasserman, and Pedro Cavaleiro Miranda

Standardizing Skullremodeling Surgery and Electrode Array
Layout to Improve Tumor Treating Fields Using Computational
Head Modeling and Finite Element Methods. 19
N. Mikic, F. Cao, F. L. Hansen, A. M. Jakobsen, A. Thielscher,
and A. R. Korshøj

Part II Low Frequency Electromagnetic Modeling and Experiment: Neural Stimulation in Gradient Coils

Peripheral Nerve Stimulation (PNS) Analysis of MRI
Head Gradient Coils with Human Body Models 39
Yihe Hua, Desmond T. B. Yeo, and Thomas K. F. Foo

Part III Low Frequency Electromagnetic Modeling and Experiment: Transcranial Magnetic Stimulation

Experimental Verification of a Computational Real-Time
Neuronavigation System for Multichannel Transcranial
Magnetic Stimulation 61
Mohammad Daneshzand, Lucia I. Navarro de Lara, Qinglei Meng,
Sergey N. Makarov, Işıl Uluç, Jyrki Ahveninen, Tommi Raij,
and Aapo Nummenmaa

**Evaluation and Comparison of Simulated Electric Field Differences
Using Three Image Segmentation Methods for TMS** 75
Tayeb Zaidi and Kyoko Fujimoto

**Angle-Tuned Coil: A Focality-Adjustable Transcranial
Magnetic Stimulator** 89
Qinglei Meng, Hedyeh Bagherzadeh, Elliot Hong, Yihong Yang,
Hanbing Lu, and Fow-Sen Choa

**Part IV   Low Frequency Electromagnetic Modeling and Experiment:
            Spinal Cord Stimulation**

**Interplay Between Electrical Conductivity of Tissues
and Position of Electrodes in Transcutaneous Spinal
Direct Current Stimulation (tsDCS)** 101
Sofia R. Fernandes, Mariana Pereira, Sherif M. Elbasiouny,
Yasin Y. Dhaher, Mamede de Carvalho, and Pedro C. Miranda

**Part V   High Frequency Electromagnetic Modeling and Experiment:
           MRI Safety with Active and Passive Implants**

**RF-induced Heating Near Active Implanted Medical Devices
in MRI: Impact of Tissue Simulating Medium** 125
James E. Brown, Paul J. Stadnik, Jeffrey A. Von Arx,
and Dirk Muessig

**Computational Tool Comprising Visible Human Project®
Based Anatomical Female CAD Model and Ansys
HFSS/Mechanical® FEM Software for Temperature Rise
Prediction Near an Orthopedic Femoral Nail Implant During a
1.5 T MRI Scan** 133
Gregory Noetscher, Peter Serano, Ara Nazarian,
and Sergey N. Makarov

**Part VI   High Frequency Electromagnetic Modeling:
            Microwave Imaging**

**Modeling and Experimental Results for Microwave Imaging
of a Hip with Emphasis on the Femoral Neck** 155
Johnathan Adams, Peter Serano, and Ara Nazarian

Index 171

Part I
Low Frequency Electromagnetic Modeling and Experiment: Tumor Treating Fields

The Impact of Scalp's Temperature in the Predicted LMiPD in the Tumor During TTFields Treatment for Glioblastoma Multiforme

Nichal Gentilal, Ariel Naveh, Tal Marciano, Zeev Bomzon, Yevgeniy Telepinsky, Yoram Wasserman, and Pedro Cavaleiro Miranda

1 Introduction

1.1 Tumor Treating Fields (TTFields)

Tumor Treating Fields (TTFields) is an anti-mitotic cancer treatment technique used for solid tumors. It consists in applying an electric field (EF) with a frequency between 100 and 500 kHz in two perpendicular directions alternately to affect the mitosis of tumoral cells [1]. TTFields are FDA-approved for the treatment of recurrent glioblastoma (GBM) since 2011, of newly diagnosed GBM cases since 2014, and of malignant pleural mesothelioma since 2019, after the promising results from the EF-11 [2], EF-14 [3] and STELLAR [4] clinical trials, respectively. The mechanisms by which TTFields act are not completely understood, but in-vitro studies showed that these EFs can slow down the rate at which tumoral cells divide or even completely arrest cell proliferation for EF intensities at the tumor of at least 1 V/cm [1, 5].

In patients, TTFields are applied using a specific device named Optune (Novocure, Haifa, Israel) which consists in an electric field generator connected to four arrays with 9 transducers each. These arrays work in pairs to induce an EF in the tumor in two perpendicular directions alternately. The rationale behind the application of the fields in more than one direction is based on the results of the in-vitro study by Kirson et al. [5] that showed an increase of 20% in the number of

N. Gentilal (✉) · P. C. Miranda
Instituto de Biofísica e Engenharia Biomédica, Faculdade de Ciências da Universidade de Lisboa, Campo Grande, Lisbon, Portugal
e-mail: ngentilal@fc.ul.pt

A. Naveh · T. Marciano · Z. Bomzon · Y. Telepinsky · Y. Wasserman
Novocure LTD, Haifa, Israel

© The Author(s) 2023
S. Makarov et al. (eds.), *Brain and Human Body Modelling 2021*,
https://doi.org/10.1007/978-3-031-15451-5_1

tumoral cells arrested during mitosis when two directions were used alternately compared to a single direction.

Post-hoc analyses of clinical trials data showed that the time that a patient is on active treatment, i.e., the treatment compliance, significantly affects the expected overall survival (OS) [6]. Patients with recurrent GBM tumors that participated in the EF-11 trial and whose daily compliance was at least 18 hours had an OS of 7.7 months, whereas those who were under treatment less than these 18 daily hours had an OS of 4.5 months. More recently, Ballo et al. [7] used data from newly diagnosed GBM patients who took part on the EF-14 clinical trial to find a relation between TTFields dose and treatment outcome. The results showed that the OS survival is significantly different in patients whose average electric field in the tumor was at least 1.06 V/cm (24.3 months vs. 21.6 months), thus corroborating what was seen in in-vitro studies [1]. In the same study, the authors suggested a new metric, the power density (PD), that could be used to evaluate treatment efficacy instead of the electric field. By defining the dose in the tumor in the terms of PD, TTFields planning will be more like other cancer treatment techniques planning, such as radiotherapy, as this metric quantifies the energy that is deposited in tissues. The same study showed that the threshold that best divides patients into the two groups with the most significant difference in terms of OS is 1.15 mW/cm^3.

1.2 Optimization of the Array Placement: The NovoTAL System

To optimize the PD in both directions, the array layouts are strategically placed on the scalp in regions that maximize the dose delivered to the tumor. The NovoTAL system (Novocure, Haifa, Israel) is an FDA-approved tool used to help certified physicians find the best array layouts for each patient. The importance of personalized array layouts is clearly shown in the literature by different authors. The study by Wenger et al. [8] was the first work to quantify the importance of creating personalized treatment maps for each patient based on tumor location. In general, the average field intensity in the tumor increased between 32 and 45% when array positioning was adjusted to the tumor location. Smaller improvements were seen in the work by Korshoej et al. [9], in which adapted layouts led to an enhancement of the average field strength in the tumor between 9% and 23%.

In the NovoTAL system personalized head models are created and several array layouts are tested to find the one that yields the best option for each patient. One of the approaches that is most commonly considered comprehends an extensive search in which a finite number of layouts, available in the NovoTAL database, are tested and the one that induces the highest EF in the tumor is chosen. After the best layout is known small adjustments can be made to optimize the induced electric field. A detailed description of how this software works can be found in [10].

1.3   Heat Transfer During Treatment

In the studies mentioned above, the choice of which layout to use was based solely on the electric field delivered to the tumor. However, there are other factors that can affect the current that is injected in each array pair during the therapy and consequently the predicted treatment efficacy. As TTFields are applied for long periods of time, the temperature of the head increases inevitably because of the Joule effect. To avoid any thermal harm to tissues, the Optune device monitors the temperature of the scalp and keeps it around 39.5 °C by controlling the amount of current that is injected into each array pair. As the EF produced at the tumor by a given pair of arrays is proportional to the injected current, a decrease in the injected current implies a reduction in the EF at the tumor.

Recent computational studies [11–13] modelled how TTFields were applied considering these thermal restrictions, quantified the temperature increases in tissues and predicted what physiological changes could occur when TTFields were applied. These works showed that heating was very localised, and it occurred mainly underneath the regions where the transducers were placed. The scalp was the tissue that heated up the most and the brain the one whose temperature varied the least. In each tissue, the temperature increased the most at the surface and quickly decreased with depth [13].

1.4   Aim of the Work

Given that the scalp's temperature limits the dose that is administrated to the tumor, and consequently the predicted treatment effectiveness, in this work we investigated the impact of including this parameter on the choice of the best layout for TTFields treatment.

2   Methods

2.1   The Simplified Head Model (SHM)

As a first step we used a simplified head model (SHM) to investigate the impact that the temperature might have in the metric used to choose the best layout. This model is composed by several spheroid shells each one representing a different tissue of interest (scalp, skull, CSF, brain and ventricles) as shown in Fig. 1. The dimensions assigned to each shell were based on the typical dimensions of the head of patients that participated in the EF-14 clinical trial. A spherical lesion with a radius of 7.1 mm was also added to the model to mimic a GBM tumor.

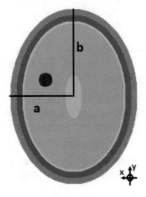

Tissue	a (mm)	b (mm)
Scalp	75	100
Skull	68	93
CSF	61	86
Brain	59	84
Ventricles	10	25

Fig. 1 Axial cut through the center of the model (plane z = 0). Each tissue was represented as a spheroid shell with the values of the main semi-axes, a and b, shown in the table on the right, in mm. The scalp is represented in orange, the skull in blue, CSF, which also fills the ventricles, in yellow and the brain in green. A spherical lesion, with the center at (34,18,0) mm and diameter of 15.2 mm, in brown, was also added to the model to mimic a GBM tumor

To maximize the number of layouts that could be tested we created several models with transducer arrays placed in different regions of the scalp. All models were created using the Materialise 3-matic software. One important consideration to bear in mind when placing the arrays is that the 2 pairs should induce EFs with directions as close as possible to being perpendicular to increase the number of affected tumoral cells [1]. One pair, hereinafter referred to as pair 0, was placed as depicted in the top row of Fig. 2, in red. The array vector (AV), which was defined as the line that connects the two central transducers of each pair, passed through the center of the tumor. The direction of the array vector is 0° when it points along the y-axis and 90° when it points along the x-axis. The perpendicular pair, pair 90, in blue in the center of the top row of Fig. 2, was rotated about the z-axis and its array vector also passed through the center of the tumor. In this work, we also considered the combinations between pair 0 and the pairs that were rotated by 75° and 105°, pair 75 and 105, respectively, as it might not always be possible to ensure a complete perpendicularity between pairs in a real-case scenario. In total, we build seven different models in which the pairs were separated by 90° ± 15° (Fig. 2). In all pairs the array vector passed through the center of the virtual lesion.

The arrays are a 3×3 matrix of transducers separated by 44 mm in one direction and by 22 mm in the other, thus representing the Optune system. The radius of each transducer is 10 mm, and its thickness is 1 mm. Between each transducer and the scalp, a layer of gel with a radius of 10.3 mm and variable thickness was added to optimize the current flow into the head.

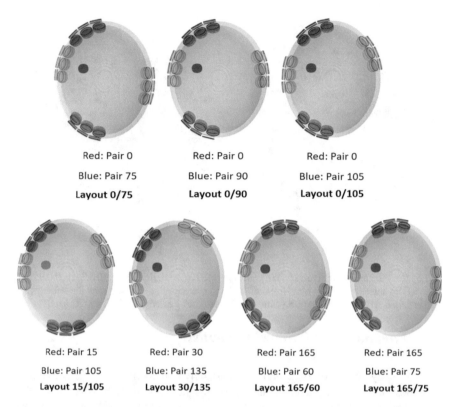

Fig. 2 Top view of the seven spheroid models used. For each pair there is a maximum of three possible complimentary pairs that can be used. For pair 0 these three pairs were rotated by 75° (upper row, left), 90° (upper row, middle) and 105° (upper row, right). For all layouts, the array vector of each pair passed through the center of the tumor, which is represented in black. The pair with the longest array vector is colored in red, whereas the pair with the shortest array vector is colored in light blue. The position of the tumor was the same for all layouts

2.2 The Realistic Head Model (RHM)

After the simulations with the simplified head model were completed, we used a realistic head model (RHM) to verify if the conclusions drawn hold for more complex geometries. The RHM was the same one used in previous studies on heat transfer during TTFields [11–13], This model was built based on the Colin27 template, and it was segmented into scalp, skull, CSF, grey matter (GM) and white matter (WM). Similar to what was done with the SHM, a spherical virtual lesion was added to mimic a GBM tumor. The latter consisted in a necrotic core, with a diameter of 14 mm, surrounded by an active tumor shell, with a diameter of 20 mm. A description of the complete framework followed to build this model can be found in [14, 15]. The arrays were then placed on the scalp using Materialise 3-matic, in positions defined by the NovoTAL system. The final head model is shown in Fig. 3.

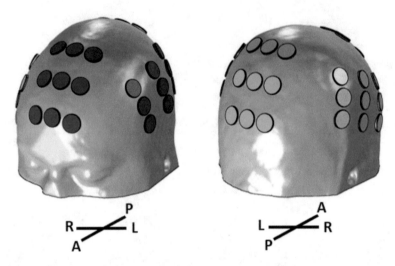

Fig. 3 Realistic head model used with the arrays placed in the regions suggested by the NovoTAL system. The anterior array (in blue) is paired with the posterior one (in green) and the left array (in red) is paired with the right one (in yellow). A: Anterior; P: Posterior; L: Left and R: Right

2.3 Equations and Physical Properties

As previously noted, current is injected alternately in two directions. Thus, to model TTFields, it is necessary to calculate the electric field distribution twice. At the frequency at which TTFields work for GBM treatment, 200 kHz, it is possible to use the electroquasistatic approximation of Maxwell's equation. In this case, the calculation of the electric potential is simplified, and Laplace's equation can be used:

$$\nabla\left[\left(\sigma + j\epsilon_a 2\pi f\right)\nabla\phi\right] = 0 \tag{1}$$

where σ is the electric conductivity (S/m), j the imaginary constant, ϵ_a the absolute permittivity (F/m), f the frequency (Hz) and ϕ the electrostatic scalar potential (V).

As Optune controls the potential difference between the two arrays of the same pair, we imposed a Dirichlet boundary condition ($V = \pm V_0$) at the outer surfaces of the transducers of the same array. At the remaining outer boundaries of the model, we assumed a null value for the normal component of the current density, whereas at the internal boundaries we considered continuity of the latter quantity. The electric field distribution was obtained for the two array pairs in each layout.

After solving the electric studies, we used Pennes' equation to solve for the temperature distribution:

$$\rho c \frac{\partial T}{\partial t} = \nabla\cdot\left(k\nabla T\right) + \omega\rho_b c_b\left(T_b - T\right) + Q_{met} + \sigma\left\|E\right\|^2 \tag{2}$$

where ρ is the density (kg/m³), c the specific heat at constant pressure (J/(kg °C)), T the temperature (°C), t the time (s), k the thermal conductivity (W/(m °C)), ω the blood perfusion (1/s), Q_{met} the metabolic rate (W/m³), assumed to be constant in the simulations, and E the electric field vector (V/m). The last term on the right-hand side represents the contribution of the EFs in increasing the temperature. Its values were taken alternately from each array pair. The subscript b stands for blood and T_b was 36.7 °C in all studies.

At the outer boundaries, we accounted for energy exchanges with the environment, at 24 °C, through convection and radiation, whereas at the internal boundaries we assumed that heat conduction occurred between adjacent tissues and materials.

The values assigned to each physical property were based on the literature. The electric parameters were retrieved from Ballo et al. [7] and the thermal parameters were the same as the standard values reported in Gentilal and Miranda [12].

All simulations were performed in COMSOL Multiphysics v5.2a.

2.4 Conditions of the Simulations and Metrics Used

As described in the work by Chaudry et al. [16], during treatment planning with NovoTAL the electric field in the tumor is predicted by injecting 900 mA of current (amplitude) into each array pair independently. Additionally, in the work by Ballo et al. [7], the local minimum power density (LMiPD) is defined as the lowest of the two power densities induced in each voxel in the tumor. Thus, we combined these two approaches and, when only the electric field was considered as a metric to evaluate the effectiveness of a layout, we injected 900 mA into each array pair, and we calculated the LMiPD in the tumor. As noted by Wenger et al. [10], it is desirable to use MR images with a resolution of at least 1 mm × 1 mm × 1 mm for TTFields planning. Thus, the values of the power density in the tumor were exported from COMSOL considering a regular grid spaced by 1 mm in every direction. After the minimum power density was obtained for each voxel, we calculated the average LMiPD in the tumor for each layout. Each electric study took around 4 minutes using the SHM and 30 minutes using the RHM in a workstation with dual core Intel Core i9-10900X X-series processors clocked at 3.7 GHz and with 64 GB of RAM.

When the temperature was also accounted for during treatment planning a different approach had to be followed as the amount of current that could be injected into each array pair depended on the temperature of the scalp. We used a method previously reported [13] in which we started by injecting 400 mA amplitude of current into each pair alternately with a switching time of one second. We then predicted the maximum temperature that the scalp would reach underneath each array pair in regions where the thermistors are typically placed, and we iteratively fine-tuned the injected current to induce the desired maximum temperature underneath both pairs.

Predicting the temporal evolution of the temperature distribution required a time-dependent study, with the electric field in Eq. (2) switching between that produced by each of the two array pairs every second. Only the first 5 minutes of treatment were simulated due to the time taken by these simulations. In order to estimate the steady state temperature, we fitted the following general equation to the variation of the highest temperature in the scalp:

$$T = C_1\left(1 - \exp\left(-t/C_2\right)\right) + C_3 \tag{3}$$

where C_1 (°C) represents the maximum contribution that the EF has in increasing the temperature of the scalp, C_2 (s) is a time constant related to how quick the heat is transferred and C_3 (°C) represents the initial temperature of the scalp.

The maximum temperature, T^{max}, can then be predicted by taking the limit as t tends to infinity:

$$T^{max} \equiv \lim\left(t \rightarrow \infty\right)T = C_1 + C_3 \tag{4}$$

All data curve fitting were performed using Matlab's 2020 curve fitting toolbox (cftool).

In the case of the RHM the desired temperature limit was the optimal working point for the Optune device, 39.5 °C. For the SHM, we had to slightly increase this threshold, to 40.5 °C, as the minimum current that is typically injected in patients, 400 mA, already induced a temperature higher than 39.5 °C. The time-dependent studies took around 10 hours for the SHM and 34 hours using the RHM in the workstation aforementioned.

3   Results and Discussion

3.1   Simplified Head Model

3.1.1   LMiPD Values When Only the Electric Field Is Considered

The values of the LMiPD when 900 mA (amplitude) were injected into each pair using the simplified head model are presented in Table 1.

The values of the LMiPD were very similar for all layouts (range: 6.30–6.88 mW/cm³). The worst layout for this tumor position, layout 165/75, induced a LMiPD that was only 8% lower than the one induced by the best layout, layout 15/105. These values were mainly limited by the lower EF induced by the pair with the longest array vector. In general, a longer array vector corresponds to a more resistive current path. However, resistance values also did not change considerably across layouts: 90–93 Ω for the pair with the highest AV and 77–82 Ω for the complimentary pair. Under these conditions, the parameter that defined which was the best layout is the distance between the tumor and each array (Fig. 2). In layout 15/105 the tumor is very close to two arrays of different pairs thereby increasing the induced PD in both

Table 1 LMiPD values in the tumor when 900 mA (amplitude) were injected into each pair. Each layout is named after the two pairs which compose it, where the first pair is the one with the longest array vector and consequently the highest resistance

Layout	Injected current (mA) First pair	Injected current (mA) Second pair	LMiPD (mW/cm³)
0/75	900	900	6.63
0/90	900	900	6.61
0/105	900	900	6.57
15/105	900	900	6.88
30/135	900	900	6.71
165/60	900	900	6.31
165/75	900	900	6.30

directions. The distance between the tumor and the closest array of pair 15 is 55 mm, a value only slightly higher when compared to the distance to the closest array of pair 105, 48 mm. On the other hand, in layout 165/75, the distance between the tumor and the closest array of pair 75 is only 36 mm, but this value increases to 78 mm when the distance to the closest array of pair 165 is considered. Thus, the LMiPD in the tumor will be higher for layout 15/105 compared to layout 165/75 for the same injected current.

Based only on the LMiPD values, the layout that would be chosen for this model is layout 15/105.

3.1.2   LMiPD Values When the Temperature Is Accounted for

To investigate how the temperature impacts treatment planning we started by injecting 400 mA into each array pair alternately with a switching time of 1 second and, at the end of the simulation, we predicted the maximum temperature that the scalp would reach underneath both pairs. We then iteratively fine-tuned the injected current to induce the highest average current, considering both pairs, that would lead to a maximum of 40.5 °C in the scalp.

Taking layout 0/75 as an example, Fig. 4 shows the maximum temperature variation of the scalp for different sets of injected currents.

The maximum temperature in the scalp for each set of injected currents was predicted using Eq. (4) and it is presented in Table 2. The minimum adjusted-R^2 (A-$R^2$2) obtained during the curve fitting process was 0.9995.

An increase of 200 mA of current in pair 75, from 400 mA to 600 mA, led to a temperature increase of 2.8 °C in the scalp underneath this pair and of 0.4 °C underneath pair 0 as the two pairs were placed very close to each other. A more detailed description of the impact of the distance between array pairs is provided in the next subsection.

The same rationale was followed for the remaining 6 layouts. The LMiPD was then quantified for the current values that led to 40.5 °C in the scalp (Table 3). The lowest A-R^2 was 0.9994.

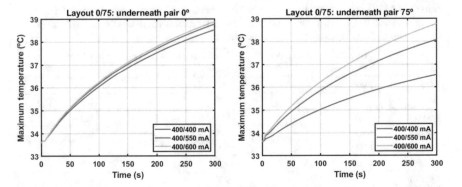

Fig. 4 Scalp's maximum temperature variation underneath pair 0 (left) and pair 75 (right) for different values of injected current. Each line corresponds to a different set of values denoted by X/Y mA, where X is the current injected in pair 0 and Y the current injected in pair 75. Current was injected alternately into each pair with a switching time of 1 second. Changing how much current was injected in pair 75 also changed the temperature of the scalp underneath pair 0

Table 2 Maximum temperature predicted in the scalp underneath each pair of arrays when different sets of current were injected in each pair alternately with a switching time of 1 second

Injected current (mA) Pair 0/Pair 75	Maximum scalp's temperature Underneath pair 0 (°C)	Maximum scalp's temperature Underneath pair 75 (°C)
400/400	40.1	37.7
400/550	40.4	39.5
400/600	40.5	40.5

Table 3 LMiPD values in the tumor for each layout when current was injected into each array pair alternately with a switching time of 1 second. In every case, the amount of current injected was iteratively refined to induce a maximum temperature of 40.5 °C in the scalp underneath each pair

Layout	Injected current (mA) First pair	Injected current (mA) Second pair	T + LMiPD (mW/cm³)
0/75	400	600	1.31
0/90	410	625	1.38
0/105	423	600	1.46
15/105	400	575	1.44
30/135	425	600	1.71
165/60	423	600	1.39
165/75	423	625	1.39

Accounting for the temperature led to a variation of 31% in the LMiPD values (range: 1.31–1.71 mW/cm³), a value higher than the 8% reported when only the electric field was used as a metric to predict treatment effectiveness. In general, the current injected into the most resistive pair did not vary by more than 25 mA (range: 400–425 mA), whereas the current injected into the complimentary pair varied between 575 and 625 mA.

Based on Table 3, the best treatment option is layout 30/135. When only the electric field was used as a criterion to choose layouts (Table 1), this layout would be the second choice, after layout 15/105. On the other hand, layout 15/105 would now be the third choice if information regarding scalp's temperature was included during treatment planning.

3.1.3   Comparison Between LMiPD and T + LMiPD

Figure 5 depicts the variation in the ranking of the layouts based on the values of the LMiPD when the temperature is not accounted for during treatment planning (LMiPD) vs. when it is (T + LMiPD).

Based on these data, it is noteworthy that layouts 0/75, 0/90 and 15/105 dropped in rank when information regarding scalp temperature was included in treatment planning. By definition, the LMiPD is limited by the pair that induces the lowest power density in the tumor. As less current is injected into the most resistive pair in these three layouts compared to the other four (first pair, Table 3), the LMiPD will be reduced. As it can be seen in Fig. 2, some pairs in these three layouts are close to each other and thus the current that is injected into one pair will not only increase the temperature in scalp regions underneath that pair but also contribute to increasing the temperature underneath the complimentary pair. This can be seen quantitively in the data presented in Table 2, where an increase of 200 mA in pair 75 led to an increase of 0.4 °C underneath pair 0.

In particular, two pairs in layouts 0/75, 15/105 and 0/90 created a temperature hotspot near the two closest transducers of different arrays, as shown in Fig. 6. In the first two layouts the shortest distance between transducers of two different pairs

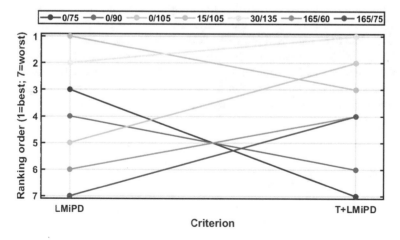

Fig. 5 Variation in the ranking of the layouts based on two different criteria: only the electric field in the tumor (LMiPD) or the electric field in the tumor plus information regarding the temperature of the scalp (T + LMiPD). Each line represents a different layout

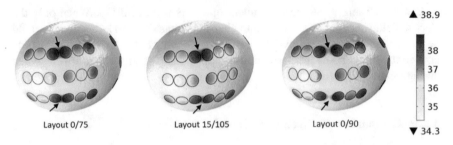

Fig. 6 Temperature distribution at the end of the simulation (t = 5 minutes) in layouts 0/75 (left), 15/105 (middle) and 0/90 (right), when 400/600 mA were injected in the first, 400/575 mA in the second and 410/625 mA in the third. Due to the proximity of some transducers of different pairs in these three layouts temperature hotspots occurred (black arrows) which limited how much current could be injected into each pair. In the first two the shortest distance between transducers of different pairs was 2.2 mm, whereas in the third it was 10.0 mm. Temperature scale is in °C

was 2.2 mm. In layout 0/90, this distance was 10.0 mm, which created a weaker but still significant hotspot.

This highlights the importance of considering scalp's temperature during TTFields planning, especially in patients with smaller heads in which there is a high probability for the transducers of different pairs being very close to each other.

The average value of the LMiPD across layouts decreased from 6.59 mW/cm³, when only the electric field in the tumor was considered, to 1.51 mW/cm³, when information concerning the temperature of the scalp was also included, a reduction of 77% on the average LMiPD predicted in the tumor. This is due to the large reduction in injected current imposed by the limit on the scalp temperature.

This simplified head model allowed us to gain important insights on some of the most important parameters when planning TTFields. However, the conclusions drawn from it might only be useful qualitatively due to some limitations such as the small variation in the values of the metrics used and the fact that the temperature threshold had to be increased from 39.5 °C to 40.5 °C. The latter might suggest that the heat mechanisms, mainly the Joule heating term, is overcontributing to the temperature increases as this threshold was reached for low values of injected current compared to what is typically seen in patients. In the next section, we present a preliminary study where we performed the same type of analysis using a realistic head model.

3.2 Realistic Head Model

3.2.1 LMiPD vs. T + LMiPD

In the realistic head model, the LMiPD in the tumor was also first calculated by injecting 900 mA into each pair, which yielded 0.81 mW/cm³. This value is below the 1.15 mW/cm³, reported by Ballo et al. [7], which is the threshold that best divides GBM patients into two groups with the most significant differences in the overall survival. One important factor limiting the LMiPD in this head model is the electric field produced by the anterior-posterior (AP) configuration. The power density in the tumor induced by this pair was 0.88 mW/cm³, a value far below the 3.32 mW/cm³ induced by the left-right (LR) configuration. The virtual lesion was placed in the right hemisphere of the brain, close to the lateral ventricles but midway between the anterior and posterior arrays, where the electric field produced by the AP pair was much lower than to the one induced by the LR pair. In general, tumors located in deeper regions of the brain, as in this case, require a more robust placement of the layouts as clearly shown in the work by Korshoej et al. [9]. According to the results of the latter work, a layout that might yield a higher LMiPD in our head model could be one in which all pairs are maintained at the same height, but rotated by 45°. This would decrease the power density induced by the LR pair but increase the one produced by the AP pair, which is the pair that is strongly limiting the LMiPD.

We then included information regarding the temperature of the scalp in the simulations. We followed the same approach as before and our results indicated that if 580 mA were injected into the AP pair and 850 mA in the LR pair alternately, the maximum temperature that the scalp would reach underneath both pairs would be 39.5 °C. The lowest A-R² obtained for the data used to drawn these conclusions was 0.998.

More current could be injected in the left-right configuration because the head is less resistive in this direction than in the AP direction. Figure 7, shows the temperature distribution in scalp's surface at the end of the simulation (t = 5 minutes). As expected, heating was very localised, and it occurred mainly underneath the regions

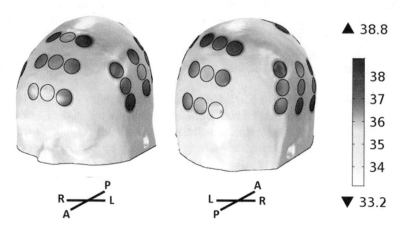

Fig. 7 Scalp surface temperature after the first 5 minutes of treatment when 580 mA of current were injected into the AP pair and 850 mA into the LR pair, alternately with a switching time of 1 second. The maximum temperature predicted in the scalp under these conditions is 39.5 °C underneath both pairs. As it can be seen, heating is very localised, and the temperature increases underneath one pair do not significantly affect the increases underneath the complimentary pair. Temperature scale is in °C

where the arrays were placed. Contrary to what occurred with layouts 0/75, 15/105 and 0/90 in the SHM (Fig. 6), in this model the two pairs were relatively far from each other, and thus common temperature hotspots did not occur.

The LMiPD was then calculated for these conditions (T + LMiPD) and compared with the values obtained when only the electric field was used as a criterion. The LMiPD induced by the AP configuration decreased from 0.88 to 0.36 mW/cm³ and the one produced by the LR pair decreased from 3.32 to 2.96 mW/cm³. The overall LMiPD predicted in the tumor decreased from 0.81 mW/cm³ to 0.36 mW/cm³, when the temperature was also included in treatment planning, a 56% reduction. This suggests that taking the scalp's temperature into account is likely to introduce large variations in the LMiPD also when using realistic head models.

4   Conclusions and Future Studies

The results obtained using the SHM model suggested that the best treatment layout in terms of LMiPD was not necessarily the best one when the temperature was included in treatment planning. In some cases, layouts in which two array pairs were close to each other yielded the highest average EF in the tumor, but the occurrence of a common temperature hotspot significantly limited the current that could be injected into each array pair, and thus the predicted treatment efficacy when the temperature was accounted for. However, adding this additional step during TTFields treatment planning might be very time-consuming. In our workstation, it took 10 hours to compute each time-dependent simulation. The process of fine-tuning the values of the injected current and of testing several layouts for each model might lead to computational times that are unfeasible from a practical point of view. Thus, further investigation is needed to include thermal studies in the treatment planning workflow. One important consideration that might be helpful is to define a minimum distance between the arrays to prevent the occurrence of temperature hotspots.

The results obtained with the RHM are a part of a preliminary study that is under investigation. The data presented here were intended to show that the LMiPD predicted in the tumor can also significantly decrease in realistic head models when the temperature of the scalp is considered during treatment. We are currently evaluating the impact of these decreases using different array layouts obtained through the NovoTAL system. In some of them, transducers of different pairs are also very close to each other, as it occurred with the SHM, and thus significant decreases in the predicted LMiPD are expected. Furthermore, we are also investigating different metrics that could be used alongside LMiPD to predict more realistically treatment effectiveness.

Acknowledgements Instituto de Biofísica e Engenharia Biomédica is supported by Fundação para a Ciência e Tecnologia (FCT), Portugal, under grant n° UIDB/00645/2020.

The authors would like to thank Oshrit Zeevi, from Novocure, for helping in the creation of the simplified head model.

Conflict of Interests Pedro Cavaleiro Miranda has a research agreement funded by Novocure. Nichal Gentilal holds a PhD grant from the research agreement with Novocure.

Ariel Naveh, Tal Marciano, Zeev Bomzon, Yevgeniy Telepinsky and Yoram Wasserman are employed by, and hold stock in, Novocure.

References

1. E.D. Kirson, Z. Gurvich, R. Schneiderman, E. Dekel, A. Itzhaki, Y. Wasserman, R. Schatzberger, Y. Palti, Disruption of cancer cell replication by alternating electric fields. Cancer Res. **64**(9), 3288–3295 (2004). https://doi.org/10.1158/0008-5472.CAN-04-0083

2. R. Stupp, E. Wong, A.A. Kanner, D. Steinberg, H. Engelhard, V. Heidecke, et al., NovoTTF-100A versus physician's choice chemotherapy in recurrent glioblastoma: a randomised phase III trial of a novel treatment modality. Eur. J. Cancer **48**(14), 2192–2202 (2012). https://doi.org/10.1016/j.ejca.2012.04.011

3. R. Stupp, S. Tailibert, A.A. Kanner, W. Read, D. Steinbergm, B. Lhermitte, et al., Effect of tumor-treating fields plus maintenance temozolomide vs maintenance temozolomide alone on survival in patients with glioblastoma. JAMA **318**(23), 2306–2316 (2017). https://doi.org/10.1001/jama.2017.18718

4. G. Ceresoli, J. Aerts, R. Dziadziuszko, R. Ramlau, S. Cedres, J. van Meerbeeck, et al., Tumour treating fields in combination with pemetrexed and cisplatin or carboplatin as first-line treatment for unresectable malignant pleural mesothelioma (STELLAR): a multicentre, single-arm phase 2 trial. Lancet Oncol. **20**(12), 1702–1709 (2019). https://doi.org/10.1016/S1470-2045(19)30532-7

5. E.D. Kirson, V. Dbaly, F. Tovarys, J. Vymazal, J. Soustiel, A. Itzhaki, et al., Alternating electric fields arrest cell proliferation in animal tumor models and human brain tumors. Proc. Natl. Acad. Sci. U. S. A. **104**(24), 10152–10157 (2007). https://doi.org/10.1073/pnas.0702916104

6. A.A. Kanner, E. Wong, J.L. Villano, Z. Ram, EF-11 investigators, Post Hoc analyses of intention-to-treat population in phase III comparison of NovoTTF-100A™ system versus best physician's choice chemotherapy. Semin. Oncol. **5**(Suppl 6), S25–S34 (2014). https://doi.org/10.1053/j.seminoncol.2014.09.008

7. M.T. Ballo, N. Urman, G. Lavy-Shahaf, J. Grewal, Z. Bomzon, S. Toms, Correlation of tumor treating fields dosimetry to survival outcomes in newly diagnosed glioblastoma: a large-scale numerical simulation-based analysis of data from the phase 3 EF-14 randomized trial. Int. J. Rad. Oncol. Biol. Phys. **104**(5), 1106–1113 (2019). https://doi.org/10.1016/j.ijrobp.2019.04.008

8. C. Wenger, R. Salvador, P. Basser, P.C. Miranda, Improving tumor treating fields treatment efficacy in patients with glioblastoma using personalized Array layouts. Int. J. Radiat. Oncol. Biol. Phys. **94**(5), 1137–1143 (2016). https://doi.org/10.1016/j.ijrobp.2015.11.042

9. A. Korshoej, F. Hansen, N. Mikic, G. Oettingen, J. Sorensen, A. Thielscher, Importance of electrode position for the distribution of tumor treating fields (TTFields) in a human brain. Identification of effective layouts through systematic analysis of array positions for multiple tumor locations. PLoS One **13**(8), e0201957 (2018). https://doi.org/10.1371/journal.pone.0201957

10. C. Wenger, P.C. Miranda, Z. Bomzon, N. Urman, E. Kirson, Y. Wasserman, Y. Palti, US 2017/0120041 A1: TTFields treatment with optimization of electrode positions on the head based on MRI-based conductivity measurements (2017)

11. N. Gentilal, R. Salvador, P.C. Miranda, Temperature control in TTFields therapy of GBM: Impact on the duty cycle and tissue temperature. Phys. Med. Biol. **64**(22), 225008 (2019). https://doi.org/10.1088/1361-6560/ab5323

12. N. Gentilal, P.C. Miranda, Heat transfer during TTFields treatment: Influence of the uncertainty of the electric and thermal parameters on the predicted temperature distribution. Comput. Methods Prog. Biomed. **196**, 105706 (2020). https://doi.org/10.1016/j.cmpb.2020.105706

13. N. Gentilal, R. Savaldor, P.C. Miranda, "A thermal study of tumor-treating fields for glioblastoma therapy" chapter 3, in *Brain and Human Body Modelling 2020 Book*, pp. 37–62. https://doi.org/10.1007/978-3-030-45623-8_3

14. P.C. Miranda, A. Mekonnen, R. Salvador, G. Ruffini, The electric field in the cortex during transcranial current stimulation. NeuroImage **70**, 48–58 (2013). https://doi.org/10.1016/j.neuroimage.2012.12.034

15. P.C. Miranda, A. Mekonnen, R. Salvador, P. Basser, Predicting the electric field distribution in the brain for the treatment of glioblastoma. Phys. Med. Biol. **59**(15), 4137–4147 (2014). https://doi.org/10.1088/0031-9155/59/15/4137

16. A. Chaudhry, L. Benson, M. Varshaver, O. Farber, U. Weinberg, E. Kirson, Y. Palti, NovoTTF™-100A system (tumor treating fields) transducer array layout planning for glioblastoma: a NovoTAL™ system user study. World J. Surg. Oncol. **13**(316) (2015). https://doi.org/10.1186/s12957-015-0722-3

Standardizing Skullremodeling Surgery and Electrode Array Layout to Improve Tumor Treating Fields Using Computational Head Modeling and Finite Element Methods

N. Mikic, F. Cao, F. L. Hansen, A. M. Jakobsen, A. Thielscher, and A. R. Korshøj

1 Introduction

Tumor Treating Fields is a well-established treatment modality for glioblastoma (GBM) [1]. One factor that influences the efficacy of TTFields therapy is the electric field strength. High electric field intensities reduce the rate of tumor cell division *in vitro* [2] and increase progression-free survival (PFS) and overall survival (OS) in GBM patients, thus demonstrating the TTFields therapy dose-response relationship [3].

N. Mikic and F. Cao contributed equally with all other contributors.

N. Mikic (✉) · A. R. Korshøj
Aarhus University Hospital, Department of Neurosurgery, Aarhus, Denmark

Aarhus University, Department of Clinical Medicine, Aarhus, Denmark
e-mail: nikmik@rm.dk

F. Cao · A. Thielscher
Technical University of Denmark, Center for Magnetic Resonance, Department of Health Technology, Lyngby, Denmark

Danish Research Centre for Magnetic Resonance, Centre for Functional and Diagnostic Imaging and Research, Copenhagen University Hospital Amager and Hvidovre, Hvidovre, Denmark

F. L. Hansen
Aarhus University Hospital, Department of Neurosurgery, Aarhus, Denmark

A. M. Jakobsen
Aarhus University Hospital, Department of 3D Printing, Aarhus, Denmark

© The Author(s) 2023
S. Makarov et al. (eds.), *Brain and Human Body Modelling 2021*,
https://doi.org/10.1007/978-3-031-15451-5_2

Korshøj et al. proposed in 2016 a novel surgical technique "skull-remodeling surgery" (SR-surgery) to increase the TTFields intensity. SR-surgery involves removing bone from the skull in the form of craniectomy, burrholes, or skull thinning. The hypothesis is by removing parts of the cranium, the resistance created by the bone is reduced and the electric field intensity is strengthened focally in the underlying tissue. By using realistic computational head models and finite element method it was concluded that SR-surgery could potentially increase the TTFields intensity by 60–70% in superficial tumors. Furthermore, it was concluded that several small burrholes would induce a greater increase in field strength than one large craniectomy per skull defect area, which is an important clinical safety consideration [4]. Computational studies also indicated that TTFields electrode array placement could be optimized when taking SR-surgery into account [5].

Feasibility of the SR-surgery concept was recently demonstrated in a small investigator-initiated, single-center, open-label phase 1 trial (OptimalTTF-1) from 2016 to 2019 with 15 first recurrent glioblastoma patients. For ethical- and risk-to-reward reasons one of the inclusion criteria was that the SR-surgery should increase the TTFields intensity by a minimum of 25%. This was validated using computational methods on patient-specific models. The median relative increase in the field intensity was 32% (range 25;59). The trial concluded that SR-surgery was safe without additional toxicity [6].

Subsequently, a randomized, comparative, multi-center, investigator-initiated interventional phase 2 trial (OptimalTTF-2) was launched in October 2020 [7]. The trial will include 84 first recurrence GBM patients randomized 1:1 to TTFields therapy with or without SR-surgery. All patients will receive physician's choice medical oncological therapy. The primary endpoint is overall survival.

To simplify and standardize the intervention, the SR-surgery procedure is conducted using a standardizing operating procedure (SOP) that is easy to understand and replicate with minimum required training.

The specifications of the SOP are given below, and the following work shows preliminary documentation for the approach.

2   Method

2.1   Standardizing the SR-Surgery Configuration

A five burrhole SR-surgery configuration was chosen as shown in Fig. 1, based on previous research indicating higher efficacy compared to complete craniectomy [4]. The general rationale behind this configuration was to induce significant field enhancement in the peritumor and tumor region without compromising patient safety and with no requirement for protective headgear.

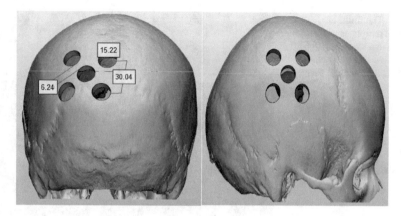

Fig. 1 Five burrholes each 15 mm in diameter. Central burrhole placed above the tumor (marked in red)

Suggested clinical procedure:

1. Five burrholes are created each 15 mm in size.
2. The central burrhole should be placed directly above the tumor resection cavity or tumor remanent.
3. Electrodes are placed so that edge electrodes from one array in a pair are located above the holes or very close to the holes. The other array in the pair should be placed directly opposite on the other side of the head. Preferably, a direct line between the transducer located above the burr-holes and the central transducer on the opposite array in the pair should pass through the region of interest, i.e., the residual tumor or a region of the resection cavity with a predicted high risk of recurrence. This configuration should be achieved from both TTFields pairs.

2.2 Technical Aspects of the SR-Surgery

To support the neurosurgeons in implementing the procedure outline above and to ensure uniformity across the phase II trial centers, 3D printed, and sterilized templates were made to guide the neurosurgeons during surgery. Four templates with different radius curvatures (0.0, 3.2, 6.4, and 9.6 mm) were included in each sterilized pack to improve fitting in different areas of the skull (Figs. 2 and 3).

The neurosurgeon will decide perioperatively, after completion of the tumor removal where the bulk of the resection cavity or tumor remanent lies and place the central burrhole directly on top. If no residual tumor is left, the surgeon would place the central burrhole above the resection cavity or a part of the surrounding brain tissue where future recurrence is expectedly likely to occur. Placement of the burrholes does not need to be restricted to the craniotomy existing bone plate used to access the tumor during resection. The template is placed as described above and

Fig. 2 The four 3D printed templates are shown in a sterilized peel-pack. Each template curves slightly more than the next

Fig. 3 The concept and schematic of a 3D printed template

the five burrholes are marked with a high-speed electrical drill. The template is then removed during drilling and repeatedly repositioned until the right configuration is achieved. The additional procedure time is approximately 5–10 min, with minimal risk of dural- and cortical damage. Two examples are given in Figs. 4 and 5.

2.3 Validating the SR-Surgery Configuration via Simulations

To validate this SR-surgery configuration we used simulations of the electric fields generated by the TTFields therapy using a detailed head model constructed from structural MR images and the Finite-Element Method (FEM). The head model was initially created from a dataset of a healthy participant which was then adapted to emulate a trial patient's pathology based on their postoperative MRI and CT of the head.

Fig. 4 A post-operative CT of the skull in a patient that underwent SR-surgery. In this case, the bone flap was big enough to fit all five burrholes while simultaneously having the central burrhole right above the resection cavity

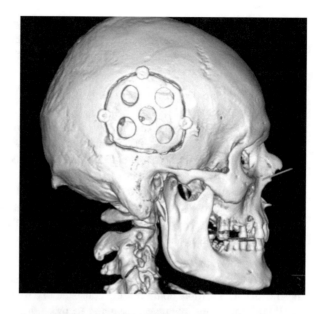

Fig. 5 Perioperative photo. It shows a bone flap that is big enough to fit all five burrholes, however, the entire configuration was moved so that the central burrhole was over the bulk of the resection cavity. Therefore a "half burrhole" was created on the bone flap and the other half on the skull

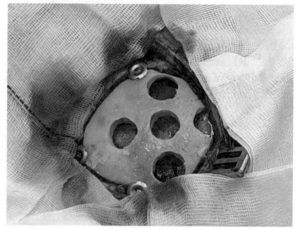

More specifically, we created the computational head model using the example dataset "Ernie" in the SimNIBS software package as a proof-of-concept demonstration. The "Ernie" dataset corresponds to a young and healthy subject, including a high-resolution T1 and a T2 weighted image. Details on the "Ernie" dataset can be found in the documentation from the SimNIBS toolbox.

In our first step, an automated tissue segmentation was performed on the T1 and T2 weighted images of the "Ernie" dataset. The initial segmentation includes the following eight tissue compartments: white matter, gray matter, CSF, scalp, skull, muscle, blood, and eyeballs.

In our second step, the segmented image was visually inspected, and the configurations of the virtual tumor resection cavity, the residual tumor, and the burrholes were set manually based on the trial patient's postoperative MRI and CT of the head and were as follows:

1. A virtual sphere-shaped tumor resection cavity with a 2.5 cm diameter in the "Ernie" head model (Fig. 6a).
2. A cylinder-shaped funnel on the top of the tumor resection cavity mimics the surgical entry to the tumor. The funnel track has a 0.8 cm diameter.
3. A sphere-shaped tumor remanent of a 2.5 cm diameter underneath the tumor resection cavity.
4. Virtual SR-surgery was applied to the head model. The SR surgery entailed placing five burrholes of 1.5 cm diameter on the skull with the central burrhole directly above the resection cavity. We investigated five different configurations of burrholes as described in Sect. 2.5.

In our third step, the head models were created using the segmented image with virtual tumor cavity, residual tumor, and with or without burrholes. The final head mesh with each burrhole configuration consisted of a total number of approximately 4,750,000 tetrahedral elements, assigned to more than 11 types, including while matter, gray matter, CSF, scalp, skull, muscle, blood, eyeballs, tumor resection cavity, residual tumor, and burrholes.

In our fourth step, the electrode arrays were placed on the skin surface for each head model (see Fig. 7) and conductivity values were assigned to the different compartments, consisting of skin (0.25 S/m), bone (0.010 S/m), CSF (1.654 S/m), grey matter (GM) (0.276 S/m), white matter (WM) (0.126 S/m), residual tumor 0.24 S/m)

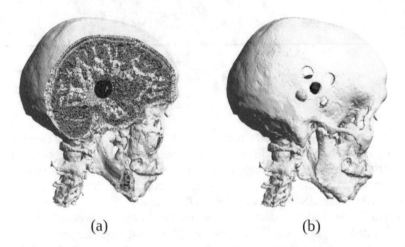

(a)   (b)

Fig. 6 (**a**) The head model is based on a healthy person "Ernie" MRI and the following spherical 2.5 cm resection cavity (blue) and spherical 2.5 cm residual tumor (yellow) was manually added based on a trial patient's postoperative MRI and CT. (**b**) shows the five burrholes, each 1.5 cm in diameter. The central burrhole is placed directly over the resection cavity

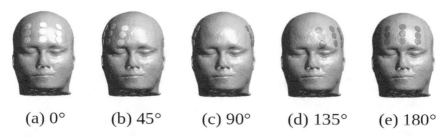

(a) 0°   (b) 45°   (c) 90°   (d) 135°   (e) 180°

Fig. 7 Surface reconstruction of the head model showing different electrode array positions used for simulation. Electrode arrays are comprised of 9 electrodes of 20 mm in diameter arranged in a 3×3 array pattern. The center-to-center distances between the electrodes are 45×22 mm. A 15-degree stepwise rotation of an electrode array was conducted around a central craniocaudal axis in the same horizontal plane, corresponding to degrees of 0, 15, 30, 45, 60, 75, 90, 105, 120, 135, 150, 165, and 180 for a total of 13 different positions. Figure (**a**)–(**e**) give five exemplary positions. Electrodes are paired gray with white

and necrotic tissue (1.0 S/m). We created a custom version of SimNIBS and provided scripts for automated simulation of TTFields induced electric fields for variations of the electrode array positions (described in the following section).

At last, the results of the simulations were visualized using Gmsh.

2.4 Electrode Array Placement

Previous research indicates that SR-surgery alters the distribution of the field density in the skull and that there are electrode array positions that are the most optimal when taking into consideration the tumor- and SR-surgery positions [5, 8].

To systematically examine the optimal placement of the electrode arrays a 15-degree stepwise rotation of an electrode array was conducted around a central craniocaudal axis in the same horizontal plane, corresponding to degrees of 0, 15, 30, 45, 60, 75, 90, 105, 120, 135, 150, 165, and 180 for a total of 13 different positions as shown in Fig. 7.

FEM calculations were performed for each electrode placement with and without the chosen SR-surgery configuration.

2.5 SR-Surgery Location in Relation to the Tumor on the Electric Field

To test the hypothesis that the burrholes should be placed directly above the tumor resection cavity to maximize the electric field directly underneath, three additional simulations were performed. Each SR-surgery configuration was moved to three alternative positions as shown in Fig. 8. The electrode arrays were rotated as

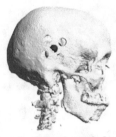

(a) Burrholes placed directly
above the cavity

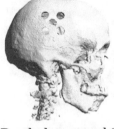

(b) Burrholes moved "up"

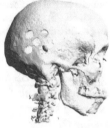

(c) Burrholes moved "back"

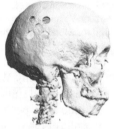

(d) Burrholes moved "up and back"

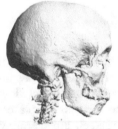

(e) Without burrholes

Fig. 8 (**a**) Top left corner shows the original placement directly above the resection cavity. The three locations were moved 3 cm in each direction "up" (**b**), "back" (**c**), and "back and up" (**d**). (**e**) is with intact skull that serves as a control

described in Fig. 7 while the electric field was calculated. For all SR-surgery, positions electrodes array rotation overlapped the burrholes. These simulations were performed to examine if small variations in the SR-surgery location had an impact on the electric field in the resection cavity and residual tumor.

3 Results

1. Burrholes significantly increased the mean, median, and peak electrical field intensities in residual tumor tissue and the resection cavity (Figs. 9 and 11).
2. The effects of burrholes on the distribution of field strengths in healthy tissues were minor (Figs. 9 and 11).
3. The highest field intensities are seen in the resection cavity and residual tumor when the burrholes and electrode array are placed as close as possible above the mentioned tissue. This is illustrated when comparing mean values of the optimal electrode array placement of 60 degrees with a suboptimal placement of 150 degrees (Figs. 10 and 13).
4. The lowest electric field values in the residual tumor tissue and resection cavity were observed for positions 0 and 180 degrees (Figs. 11, 12, and 13). The mean range for those two electrode positions was for all tissues and SR-surgery positions between 77.93 – 101.96 v/m. This indicates that electrode array positions that are far away or parallel with the burrholes should not be used since the direction of the current does not pass as strongly through the region of interest, i.e., residual tumor and resection cavity.
5. Placing the burrholes directly above the resection cavity increases the field strength in the residual tumor and resection cavity by 60–80% for most electrode array positions. The highest increase in the residual tumor was observed with the burrholes located at position (d) "up and back" and electrode placement 135–150 degrees. This aligns with the burrholes and electrode array being close to directly on top over the residual tumor but not the resection cavity (Fig. 12).

4 Summary and Discussion

The five burrhole SR-surgery configuration was chosen for OptimalTTF-2 trial, where the central burrhole should be placed above the center of the residual tumor/ resection cavity.

After the first trial patient underwent SR-surgery, the presented realistic computational head model study was performed, mimicking the patient's tumor resection cavity, residual tumor, and location. The exact burrhole location and size were mimicked from the postoperative CT scan of the head.

The field strength was calculated with and without SR-surgery including electrode array rotation around the craniocaudal axis. The results indicated this SR-surgery configuration significantly increases the electric field strength in the residual tumor and resection cavity when the electrodes are placed near or directly above the burrholes.

Finally, the SR-surgery configuration was moved to three alternative locations to examine if the central burrhole needed to be placed above the resection cavity as

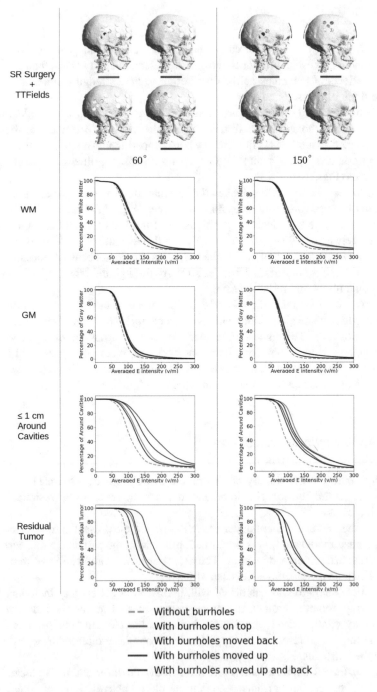

Fig. 9 Cumulative distributions of electrical field strengths were obtained with burrholes directly above the resection cavity (red lines), the three alternative placements as shown in Fig. 8, and without burrholes (dotted lines). The cumulative distributions in the y-axis are given as the percentages of tissue (white matter, gray matter, resection cavity, and residual tumor) exposed to field strengths above the corresponding values in the x-axis. The position of the electrode array is selected to compare the optimal 60-degree placement and suboptimal 150-degree placement

Fig. 10 Axial view of TTFields intensity distributions in a head model that mimicked a trial patient's postoperative MRI. The approximate positions of electrodes are shown as the surrounding patches. The left and right columns give the field distributions without and with burrholes, respectively. This show the field intensity distributions induced by electrode array positions: 0°, 45°, 90°, 135°, and 180° (their corresponding electrode montages are given in Fig. 7). The field strength increases significantly when the electrodes are placed close to or directly above the burrholes

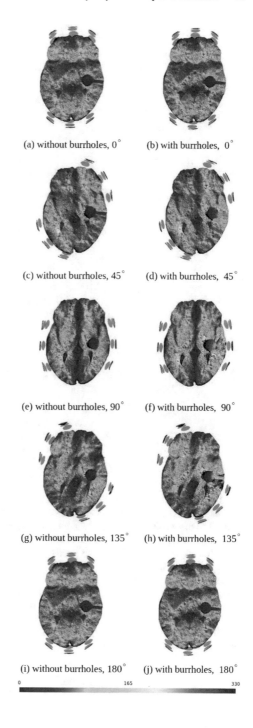

(a) without burrholes, 0°   (b) with burrholes, 0°

(c) without burrholes, 45°   (d) with burrholes, 45°

(e) without burrholes, 90°   (f) with burrholes, 90°

(g) without burrholes, 135°   (h) with burrholes, 135°

(i) without burrholes, 180°   (j) with burrholes, 180°

0   165   330

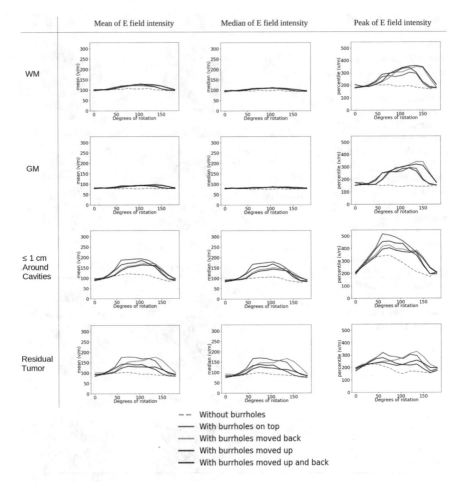

Fig. 11 Statistical analysis of TTFields intensities with burrholes directly above the resection cavity (red lines), the three alternative positions as shown in Fig. 8 and without burrholes (dotted lines). The three columns represent the mean, median, and peak values of TTFields intensities for the four tissue types (WM, GM, residual tumor, and resection cavity). The x-axis shows the degree of rotation as seen in Fig. 7 and the y-axis shows the field strength in v/m. The peak values are defined as the 99% percentile of the field intensities. The median and peak fields follow similar patterns for each tissue The TTFields intensities of WM and GM are similar with or without burrholes. The residual tumor and resection cavity field strength increase with burrholes, especially when they are close to or directly on top of the burrholes (corresponding to 60-150 degrees). Within this interval, the exact location of the electrodes can largely affect the peak values of the field intensities. The medians of field intensities are less sensitive to the electrodes' locations if they are set close to the burr holes

initiated planned. For each alternative location, the rotation of the electrode array placement around the craniocaudal axis was performed.

This concluded that the central burrhole and the entire SR-surgery configuration should be placed directly above the resection cavity and residual tumor since the

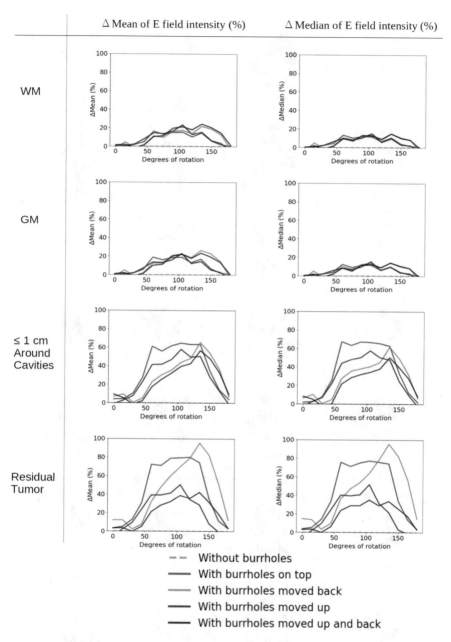

Fig. 12 The mean and median E field intensity for WM, GM, resection cavity, and residual tumor. Without burrholes are shown as the dotted line and it can be seen at the level of the x-axis. X-axis also shows the degree of electrode array location. The y-axis represents the % increase in field intensity when SR-surgery is performed. The burrholes directly on top (red lines) and their alternative locations

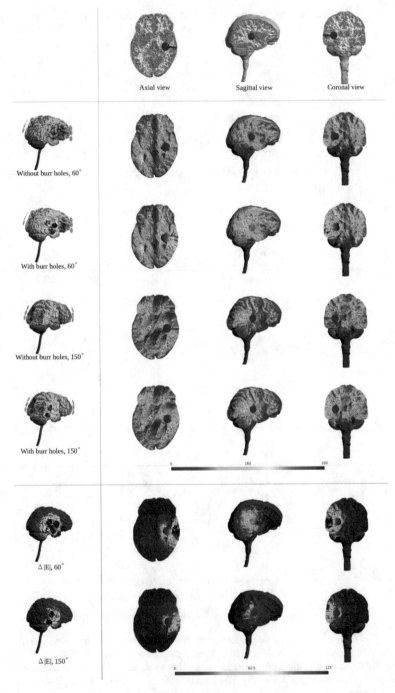

Fig. 13 Computational head model based on a healthy MRI with manually implemented features that mimicked a real patient's postoperative MRI. Burrholes were manually added based on the

burrholes significantly increase the field strength of the pathological tissue directly underneath.

These results that SR-surgery can improve the electric field and that it is possible to optimize the electrode array placement aligns well with previous research on this topic [5, 9, 10].

These results will form the basis for the standard operating procedure for SR-surgery configuration, its location, and electrode placement for each patient individually in the OptimalTTF-2 trial.

There are a number of limitations to the work presented. This study was based on one computational head model study. The results and conclusion are therefore based on a limited amount of data that might not be valid if additional simulations were made. The residual tumor and resection cavity used in the head model might not accurately reflect the complex and heterogeneous morphology and location of all glioblastoma tumors.

The five burrhole configuration was not tested in detail. Korshøj et al. conclude that several small burrholes are more effective than one big craniectomy, however, it remains unanswered what is the optimal amount of burrholes regarding field strength and without compromising patient safety.

Furthermore, the results that the burrhole location should be placed above the pathological tissue seem intuitive. In scenarios where you only have a resection cavity or a complete tumor intact, the chose where to place the burrholes will be simple. The burrhole placement choice is more uncertain when the choice is between a large residual tumor in the far periphery of the resection cavity.

Therefore, future studies are needed to validate our findings in a larger number of heterogeneous models. We plan on examining the field strength for all the trial patients with and without SR-surgery, with realistic head modeling based on their MRI, CT, electrode placement, TTFields therapy data and compare with PFS, OS, and topographical areas of recurrence seen on the patients control MRI.

Another point worth addressing is that the burrholes conductivity values were set to be the same as CSF, which may not accurately reflect reality, in which scar tissue may also constitute a part of this volume. This could give misleading results for

---

Fig. 13 (continued) postoperative CT-scan as shown in Fig. 4. The top three views (axial, sagittal) show a "funnel" or surgical access way and a resection cavity represented in blue. The tumor remnant is shown in yellow. The highest field strength was observed with electrode placement of 60 degrees. This was compared with a suboptimal electrode placement of 150-degrees. The observation is that the electric field strength is significantly enhanced in the tumor and resection cavity when electrode arrays are placed near or on the SR-surgery. **Values: 60-degree electrode array placement**. Burrholes directive above: Resection cavity 186.73 v/m, residual tumor 175.07 v/m, white matter 118.95 v/m, gray matter 91.96 v/m. Burrholes furthest away in the "back and up" position: Resection cavity 136.01 v/m, residual tumor 123.05 v/m, white matter 113.13 v/m, and gray matter 85.14 v/m. **150-degree electrode array placement**. Burrholes directly above: Resection cavity 126.49 v/m, residual tumor 118.99 v/m, white matter 105.11 v/m, and gray matter 83.25 v/m. Burrholes furthest away in the "back and up" position: Resection cavity 183.97 v/m, residual tumor 115.6 v/m, white matter 199.33 v/m, and gray matter 92.93 v/m

patients wearing TTFields therapy for months and years since their burrholes are covered in scar tissue.

More research is warranted to address several of the above-mentioned issues.

5 Conclusion

In this chapter, we continue to explore the novel idea of skullremodeling surgery to optimize TTFields therapy and the importance of electrode array placement. The concept of SR-surgery (craniectomy, burrholes, and skull thinning) and optimal electrode array placement were examined in a phase 1 pilot trial, OptimalTTF-1 that concluded skullremodeling surgery to be safe and indicated an increased survival for first recurrence glioblastoma.

The additional knowledge presented in this chapter is how skullremodeling surgery was attempted to be standardized to one SR-surgery configuration, it's placement and electrode array placement for the entire subsequent multi-center efficacy phase 2 trial "OptimalTTF-2".

Validating one SR-surgery configuration as the optimal would have a significant clinical impact since it would ensure uniformity of SR-surgery across the multi-center trial and thus increase the quality of the trial.

A five burrhole skullremodeling surgery was proposed and with one realistic computational head modeling calculated to have a significant increase effect on the residual tumor and resection cavity region without increasing the field strength for grey- and white matter. The realistic head model was recreated using patient-specific MRI and CT data. There was a significant increase in electric field strength when the electrodes were placed near the skullremodeling surgery. The skullremodeling surgery technical aspect was described. The location of skullremodeling surgery mattered and the most optimal location was with the central burrhole placed above the residual tumor or resection cavity. The data generated from these simulations will be used for the standard operating procedure for skullremodeling surgery and TTFields electrode array placement, however, the results are based on limited data. Therefore, more research is planned by validating this approach from all individual patient data gathered from the trial as it progresses and depending on these results the standard operating procedure should be updated accordingly.

References

1. R. Stupp et al., Effect of tumor-treating fields plus maintenance temozolomide vs maintenance temozolomide alone on survival in patients with glioblastoma: a randomized clinical trial. JAMA **318**(23), 2306–2316 (2017)
2. E.D. Kirson et al., Alternating electric fields arrest cell proliferation in animal tumor models and human brain tumors. Proc. Natl. Acad. Sci. U. S. A. **104**(24), 10152–10157 (2007)

3. M.T. Ballo et al., Correlation of tumor treating fields dosimetry to survival outcomes in newly diagnosed glioblastoma: a large-scale numerical simulation-based analysis of data from the phase 3 EF-14 randomized trial. Int. J. Radiat. Oncol. Biol. Phys. **104**(5), 1106–1113 (2019)
4. A.R. Korshoej et al., Enhancing predicted efficacy of tumor treating fields therapy of glioblastoma using targeted surgical craniectomy: a computer modeling study. PLoS One **11**(10), e0164051 (2016)
5. A.R. Korshoej et al., Importance of electrode position for the distribution of tumor treating fields (TTFields) in a human brain. Identification of effective layouts through systematic analysis of array positions for multiple tumor locations. PLoS One **13**(8), e0201957 (2018)
6. A.R. Korshoej et al., OptimalTTF-1: Enhancing tumor treating fields therapy with skull remodeling surgery. A clinical phase I trial in adult recurrent glioblastoma. Neurooncol. Adv. **2**(1), vdaa121 (2020)
7. N. Mikic et al., Study protocol for OptimalTTF-2: enhancing tumor treating fields with skull remodeling surgery for first recurrence glioblastoma: a phase 2, multi-center, randomized, prospective, interventional trial. BMC Cancer **21**(1), 1010 (2021)
8. A.R. Korshoej et al., EXTH-04. Guiding principles for predicting the distribution of tumor treating fields in a human brain: a computer modeling study investigating the impact of tumor position, conductivity distribution and tissue homogeneity. Neuro-Oncology **19**(Suppl 6), vi73 (2017)
9. A.R. Korshoej et al., Impact of tumor position, conductivity distribution and tissue homogeneity on the distribution of tumor treating fields in a human brain: a computer modeling study. PLoS One **12**(6), e0179214 (2017)
10. A.R. Korshoej et al., Enhancing tumor treating fields therapy with skull-remodeling surgery. The role of finite element methods in surgery planning. Ann. Int. Conf. IEEE Eng. Med. Biol. Soc. **2019**, 6995–6997 (2019)

Part II
Low Frequency Electromagnetic Modeling and Experiment: Neural Stimulation in Gradient Coils

Peripheral Nerve Stimulation (PNS) Analysis of MRI Head Gradient Coils with Human Body Models

Yihe Hua, Desmond T. B. Yeo, and Thomas K. F. Foo

1 Introduction

With the recent advancements in high-performance head gradient coil technology for magnetic resonance imaging (MRI), the assessment of peripheral nerve stimulation (PNS) has become increasingly important. In this chapter, we first briefly describe the basic concepts of PNS and gradient coils in an MRI system. We then describe and compare the significance of a head-only MRI system with respect to a whole-body system. We will also provide a general outline of the electromagnetic simulation, the neurodynamic simulation and the gradient coil design processes. In the Methods section, we deliberate the shortcomings of a widely used human body model in the head gradient coil PNS analysis and describe the modifications made to two human body models to improve the correlation of simulation and measurement results from two head gradient coils. After that, we apply these new human body models to analyze three folded and non-folded gradient coils and reveal the relationship between the eddy current flow in the human body and the gradient coil wire pattern and its impact on PNS. We also discuss the connection between PNS and concomitant fields, and the effects that a human body model with simplified tissue properties has on the PNS calculations. Finally, we will discuss the shortcomings of this study and future work and provide our conclusions.

Y. Hua (✉) · D. T. B. Yeo · T. K. F. Foo
GE Global Research, Niskayuna, NY, USA
e-mail: yihe.hua@ge.com; yeot@ge.com; Thomas.foo@ge.com

© The Author(s) 2023
S. Makarov et al. (eds.), *Brain and Human Body Modelling 2021*,
https://doi.org/10.1007/978-3-031-15451-5_3

1.1   MRI Gradient Coil and E-Field

An MRI scanner consists of several key hardware components, including the magnet, the gradient coils and the RF coils. In a typical MRI scanner, the magnet provides a homogenous magnetic field, usually 1.5 Tesla or 3 Tesla, in a field of view (FOV) that spans 45 ~ 50 cm in diameter [1]. The gradient coil system has 3 sub-coils: x, y and z. Each coil provides a linear B_z field along one axis, such that $B_{z,k} = G_k k$ ($k = x, y, z$) [2], where G_k is the gradient strength of each sub coil. The RF transmit coil transmits RF waves at the Larmor frequency of the hydrogen nucleus to excite proton spins [2] while the RF receive coil measures the resultant MR signal from these spins.

The resonance frequency $\omega = \gamma B_z$ is proportional to the applied static magnetic field, where γ is the gyro-magnetic ratio. The locations of the spins in space can be encoded by the B_z field using a combination of the main static field (B_0) and the gradient by $B_z = B_0 + \sum_{3}^{k=1} G_k k$. In a modern MRI system, different gradient waveforms are applied to each gradient sub-coil to spatially encode spins such that they have spatially varying frequencies and phases. The image of the subject can then be reconstructed from the received RF signals by utilizing image reconstruction algorithms [3].

The alternating magnetic field $\vec{B}$ generated by a gradient coil can induce an electric field $\vec{E}$ in our body, which can be calculated from the vector potential $\vec{A}$ and the scalar potential φ as $\vec{E} = -\partial \vec{A} / \partial t - \nabla \varphi$, with $\vec{A}(\vec{r}) = \dfrac{\mu_0 I}{4\pi} \int_l 1/|\vec{r} - \vec{r}'| d\vec{l}(\vec{r}')$, and t, I, $d\vec{l}$, $\vec{r}'$, $\vec{r}$ being time, current in the gradient coil, differential element in length direction, source vector and observation vector, respectively. The quantity φ in the body can be obtained from $\nabla \cdot \sigma \nabla \varphi + \nabla \cdot \sigma \partial \vec{A} / \partial t = 0$ with $\vec{j} = \sigma \vec{E}$ and the low-frequency approximation $\nabla \cdot \vec{j} = 0$, where σ and $\vec{j}$ are electric conductivity and current density, respectively. Together with the boundary condition $\dfrac{\partial \varphi}{\partial n} + \hat{n} \cdot \dfrac{\partial \vec{A}}{\partial t} = 0$ from $\hat{n} \cdot \vec{J} = 0$, $\hat{n}$ being the unit normal vector at the external surface, φ can be finally computed with numerical methods such as Finite Element Method (FEM) for complicated human body geometry with heterogenous electrical conductivity for different tissues [4]. One of the main safety-related bioeffects induced by this E-field is the stimulation of our peripheral nerves, which may cause tingling or even painful muscle contractions in severe cases. Hence, PNS is an important safety topic in gradient coil design. All gradient EM simulations in this study were performed using Sim4life Ver.6.0 (Zurich Med Tech, Zurich, Switzerland).

1.2   Head vs. Whole-Body Gradient

A very rough approximation of the E-field can give us some sense of the amplitude of the E-field. From $\nabla \times \bar{E} = -\partial \bar{B} / \partial t$ we get $\oint_l \bar{E} \cdot d\bar{l} = \dfrac{\partial}{\partial t} \int_A \bar{B} \cdot d\bar{s}$, where l is the integration path and the A is the area surrounded by the path, then [5]

$$E(r) \sim \frac{r \partial B}{2 \partial t} \tag{1}$$

It depends on the rate of change of the B-field and the volume (presented by r) to provide such a magnetic field.

A gradient coil is composed of many turns of wire as shown in Fig. 6 of reference [6]. Therefore, the gradient coil has a non-negligible inductance, which indicates that the current through it cannot change instantly. This means that $\partial B / \partial t$ and the gradient slew rate (SR), denoted by $\partial G / \partial t$ (subscript of G is removed later without loss of generality), cannot be infinite. A clinical whole-body gradient coil typically has a large FOV, resulting in a large-inductance coil that cannot support gradient waveforms with high slew rates. On the other hand, ultra-high performance head-only gradient coils designed for microstructure and functional imaging of the brain need a much smaller FOV that encompasses the head but require fast-changing gradient fields and high gradient waveform amplitudes. More importantly, head gradient coils with a smaller FOV will generate less E-field and thus have lower PNS risks.

We recently developed two high-performance head gradient systems HG2 [7] and MAGNUS [8] for research purposes. The former is designed for a compact 3 T head MRI scanner and the latter is a head gradient that can be inserted into a clinical 3 T whole-body scanner. Compared to typical clinical whole-body MRI scanners, whose gradient strength and slew rate are usually around 30 ~ 50 mT/m and 100 ~ 200 T/m/s, HG2 and MAGNUS achieve 85 mT/m, 700 T/m/s and 200mT/m, 500 T/m/s, respectively. In this study, these two coils are used in PNS model verification since we have measured data from volunteer scans.

Figure 1 shows the E-field intensity of MAGNUS and a typical whole-body gradient coil with a head scan position in the Yoon-sun human body model (v4.0b03, IT'IS Foundation, Zurich, Switzerland). It can be clearly seen that the high $|\bar{E}|$ -field region of the whole-body gradient is much larger than that of the head gradient. The high field regions by MAGNUS X and Y are mainly in the head.

1.3   Peripheral Nerve Stimulation

Nerve stimulation means an action potential is initiated and propagates through the nerve tissue. The theory is that the opening of the ion channels within the membrane of a neuron requires work from a mechanical impulse, which is further produced by

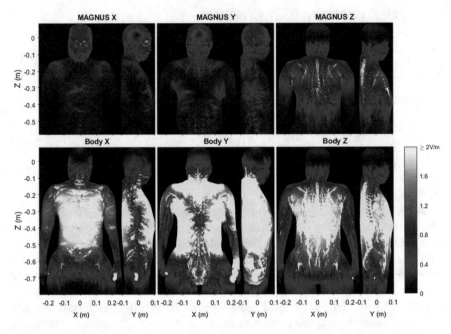

Fig. 1 E-field intensity of MAGNUS and common whole-body gradient coil with a head scan position in Yoon-sun human body model (SR is normalized to 100 T/m/s)

an electric field through time. The fundamental law for nerve stimulation by an external electric field can be expressed as [5].

$$\int_{\tau} E(t)\,dt \geq E_r\left(\tau_c + \tau\right) \tag{2}$$

where E_r is the rheobase, the minimum E-field to activate the nerve while τ_c is the chronaxie, the duration with which the threshold is exactly twice of the rheobase. If the external field is less than the rheobase E_r, no stimulation can happen even if the field is applied for an infinite time.

When the electric field is induced by the magnetic field, it is possible to use the magnetic field time derivative to describe the fundamental law. Since G and the SR $\partial G/\partial t$, are two important performance parameters for every gradient coil, the equation can be further derived as a relationship between gradient strength, slew rate and the gradient pulse duration τ [9].

$$\Delta G\left(\tau, \bar{r}\right) \geq \Delta G_{\min}\left(\bar{r}\right) + \dot{G}_r\left(\bar{r}\right)\tau \tag{3}$$

Because the E-field from the gradient coil for a certain scan position is diverse inside of the human body, the PNS response from each nerve is also different. The nerve that is easiest to be stimulated gives the threshold under this certain excitation condition.

1.4   Neurodynamic Simulation

To study the action potential initiation and propagation in detail, a closer look at the behavior of the neuron membrane is needed. The sodium and potassium gates, etc., on the neuron membrane, can be simplified as a set of voltage-controlled conductance, while other characteristics of the neuron can also be described as a set of distributed capacitance and conductance. A widely accepted computation model of myelinated mammalian nerve fibres is the MRG model [10], which utilizes double-cable RC-circuits to describe the Ranvier node, paranodal, internodal section of the axon and the myelin sheath of the axon. As a result, the action potential behavior can be simplified and described by the cable theory equation [11].

$$C_m \frac{dV_n}{dt} + G_m V_n - G_a \left(V_{n-1} - 2V_n + V_{n+1} \right) = G_a \left(V_{e,n-1} - 2V_{e,n} + V_{e,n+1} \right) \quad (4)$$

where G_m, C_m and G_a are membrane conductance, membrane capacitance and axial internodal conductance, respectively. G_m is further controlled by membrane potential reflecting the transient on-off status and the current-conducting ability of the sodium channel, potassium channel and other paths on the membrane. $V_n(t)$ and $V_e(t)$ are membrane potential response and applied potential by the external E-field, respectively (time dependency is omitted for simplicity in the equation). The excitation item (driving function) on the right side of the equation is about the second order difference of the electric potential, but actually relates the first order difference of the tangential E field along the nerve, where n is the index of the Ranvier node. With the E-field calculated from the electromagnetic simulation in the previous step, the voltage drop V_e along the nerve trajectory can be obtained by $V_e = \int_l \bar{E} \cdot d\bar{l}$. After multiplying with the time domain modulation function, $\alpha(t)$

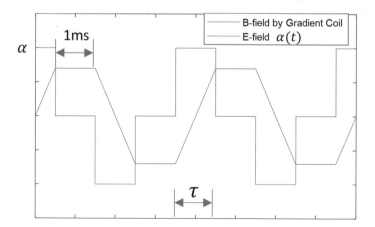

Fig. 2 Curves to represent the relationship between the B-field waveform and E-field waveform

Table 1 Parameters used in neuron simulation

Neuron model	MRG
Diameter(um)	16 (motor)
	12 (sensory)
B ramp time τ (ms)	0.2,0.5
B plateau time (ms)	1
Train length of pulses	16

(E-field waveform through time, refer to Fig. 2), and an additional scaling coefficient T,

$$V_e(t) = V_e \cdot \alpha(t) \cdot T \tag{5}$$

is finally fed into the Eq. (4) to calculate the membrane potential response for each nerve trajectory. The solver will iteratively change the scaling factor T to find the smallest field intensity to initiate an action potential. This is called the titration process. In this study, we use the NEURON module (Yale University, New Haven, CT) that is integrated in Sim4life for PNS calculations. Thus, ΔG can be calculated for every τ. Since the relationship between ΔG and τ is linear, only two values of τ are used in the simulation. Table 1 shows the parameters used for neuron simulation in Sim4Life.

1.5 Gradient Coil Design

Modern gradient coil design can be regarded as an optimization problem that seeks the optimal current distribution on a pre-defined design surface that produces a target field distribution while subjected to a set of optimization constraints [12, 13]. The current density can be decomposed into basis functions with unknown coefficients.

$$\vec{J} = \sum c_i \vec{f}_i, i = 1,\ldots,n \tag{6}$$

and the interested physical quantities such as magnetic energy and the B_z field in FOV can be expressed as quadratic form or linear form of the unknown coefficients, respectively

$$E_B = \frac{1}{2}LI^2 = \frac{1}{2}\iint_{s\,s} \frac{\vec{J}(\vec{r}) \cdot \vec{J}(\vec{r}')}{|\vec{r} - \vec{r}'|^2} dr dr' = \frac{1}{2}c^T \mathcal{L} c \tag{7}$$

$$B_z(\vec{r}) = \frac{\mu_0}{4\pi}\left(\int_S \frac{\vec{J}(\vec{r}') dr' \times (\vec{r} - \vec{r}')}{|\vec{r} - \vec{r}'|^3}\right)_z = B_z c \tag{8}$$

where $\mathcal{L} = [\mathcal{L}_{ij}]$, $\mathcal{B}_z = [\mathcal{B}_{zj}]$ and

$$\mathcal{L}_{ij} = \iint_{S \, S} \frac{\vec{f}_i(\vec{r}) \cdot \vec{f}_j(\vec{r}')}{\left|\vec{r} - \vec{r}'\right|^2} dr dr'$$

$$\mathcal{B}_{zj} = \frac{\mu_0}{4\pi} \left(\int_S \frac{\vec{f}_j(\vec{r}') dr' \times (\vec{r} - \vec{r}')}{\left|\vec{r} - \vec{r}'\right|^3} \right)_z$$

for $i, j = 1, \ldots, n$. Finally, the problem usually could be solved by a quadratic programming algorithm:

To find: $c = [c_1, c_2, \ldots, c_n]^T$

Minimizing: $\frac{1}{2} c^T \mathcal{L} c$ + other quadratic forms of the unknown coefficients...

Gradient constraint: $G_x x - \varepsilon \leq \mathcal{B}_z c \leq G_x x + \varepsilon$ for each point in X coil FOV, ε is the tolerance.

Other constraints such as force, torque, eddy current and fringe field having linear form of the unknown coefficients can also be included.

2   Methods

2.1   Augmentation of Nerves in Yoon-Sun and Duke Models

We used Yoon-sun female model v4.0b03 and Duke male model v3.1b01 from IT'IS foundation in this study. The Yoon-sun model is reconstructed from high-resolution cryosection images and it has a relatively complete nerve trajectory atlas for further neurodynamic simulation while the Duke model is based on MRI data of a volunteer without any nerve trajectories. In our previous study of PNS response for MAGNUS [14], we found that the calculated PNS thresholds for the X and Y coils are higher than those in the measured data. We hypothesized that this was because the existing nerve trajectories in the head of the Yoon-sun model are located mainly in the deeper recesses of the skull, which are in the low gradient field regions. The Yoon-sun model lacks extracranial nerves that are located in the high gradient field region.

To address this, we added extracranial nerves and superficial nerves in the neck and shoulder region to both Yoon-sun and Duke based on our knowledge of the high field area of the head gradient coils and the anatomy of the body (Fig. 3). The reason we only added a portion of the nerves is that the human body and the X-coil are mainly left-right symmetric, and it is time-consuming to add anatomically realistic nerve trajectories into the existing human body model. Table 2 summarizes the nerve branches that have been added to the human body model. The nerve branches

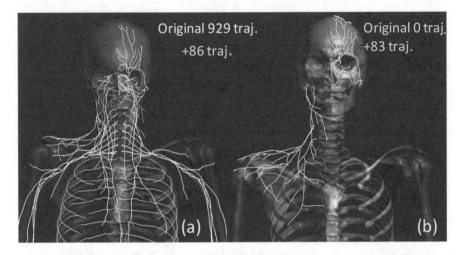

Fig. 3 (**a**) Yoon-sun: Original nerve trajectories (White), 86 nerve trajectories (Yellow) added and (**b**) Duke: 83 nerve trajectories added

Table 2 Added nerve branch names

Nerve name
Buccal branch of left facial
Temporal branch of left facial
Lateral branch of left supraorbital
Medial branch of left supraorbital
Left supratrochlear
Left infratrochlear
Left recurrent laryngeal
Right recurrent laryngeal
Right dorsal scapular
Right medial supraclavicular
Right accessory
Cervical branch of right facial

selected in Duke differ slightly from Yoon-sun based on our early calculations on the latter by omitting nerves that have a low likelihood of being stimulated.

2.2   Non-folded and Folded Gradient Coil Design

We used the gradient specifications (Table 3) from Tang's paper [12] to design three different gradient coils for this study (Fig. 4). Coil A is a non-folded design, in which the primary layer is separated from the shield layer. The second design is

Table 3 Parameters for designing folded and non-folded gradient coils

$L_{primary}$ (cm)	50.4	D_{FOV} (cm)	24
L_{shield} (cm)	72.6	Offset of FOV center to coil geometry center (cm)	−9.2
L_{TS} (cm)	137	Current (A)	400
$D_{primary}$ (cm)	33.6	Gradient strength (mT/m)	60
D_{shield} (cm)	55.8	ε (max field error in FOV)	0.1
D_{TS} (cm)	90.6	λ (max field error by eddy current)	0.002

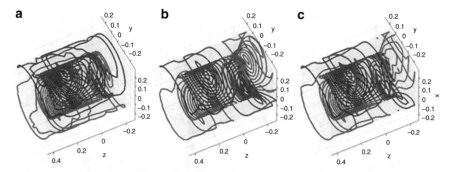

Fig. 4 Three X coils designed for this study. (**a**) A Non-folded design; (**b**) a folded design; and (**c**) a folded design with turn number limit in primary/shield connecting region

folded at the patient end, allowing more turns close to FOV which makes it more efficient. However, the wire turns that are close to the FOV are also close to the human body, which increases the E-field exposure. The third coil is also folded, but the coil turn number in the primary/shield connecting region is reduced to about half of the second coil. The idea is to seek a balance between the first and second design, and to examine how the wire pattern strategy affects the PNS in the human body.

It needs to be noted that PNS constraints were not applied in the gradient coil optimization process in this study. Relatively mature algorithms have been developed during recent years and an interested audience could read Davids' work [15] for further information. In this work, however, we mainly focus on the mechanism of how the wire pattern changes can affect the E-field, and thus, PNS.

3 Results and Discussions

3.1 Model Verification on MAGNUS and HG2 X Coils

The simulated PNS threshold by MAGNUS and HG2 X coils are within the observed range and have shown much better agreement with the measurement than before (Fig. 5a) [14]. Figure 6 shows the most sensitive places for PNS in Yoon-sun and

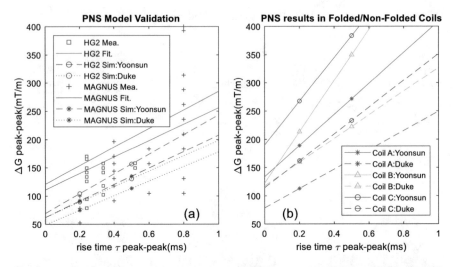

Fig. 5 (**a**) PNS simulation results compared to measurements on MAGNUS and HG2; (**b**) PNS simulation results in three folded and non-folded gradient coils

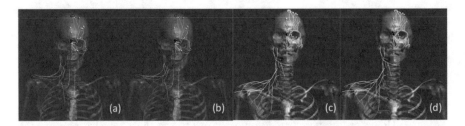

Fig. 6 PNS initiation places on nerve trajectories stimulated by MAGNUS (**a, c**) and HG2 (**b, d**) are represented by small balls. (**a, b**) are with Yoon-sun models (**c, d**) are with Duke model. The bigger and the redder of the ball means the location is more sensitive

Duke. Overall, MAGNUS and HG2 X coils have very similar PNS performances. PNS was observed to be strongest at the nasion/glabella region and also occurs with high probabilities in regions close to the eye, forehead and cheeks [16, 17]. These places are obviously different from those in [14] since there were no such nerve trajectories in the original Yoon-sun models. PNS threshold is lower in Duke than in Yoon-sun, mainly because the size of Duke is larger than Yoon-sun, and as such, the nerves in Duke can extend to higher E-field regions than in Yoon-sun. All these results are consistent with observations from volunteer scans that were reported in our previous work [16].

We also observed that simulated thresholds are lower than the measurements. It is probably because, in simulations, the action potentials are initiated mostly at the ends of the nerve trajectories, which seem to be induced by the rather abrupt termination of the nerve trajectories that is unreal. In reality, these nerves may extend longer and become narrower in diameter, having telodendria or dendrites at the ends. We have reported this effect in our previous study [14], but for comparison purposes, the simulation result is still close to measurement.

3.2 B_z and $|E|$ Field in Free Space and Human Body for Folded/Non-folded X Coils

The Bz and E-field distributions in free space in the three gradient coil designs are plotted in Fig. 7. While the B-field will be largely unperturbed by the presence of a human body model, the E-field distribution could be very different when a body model is introduced into the simulations. It could be seen that for the three designs, although the Bz fields are similar in FOV, the E-field in free space and the human body could be different. Coil A generates a relatively large E-field in the head region, while coil B results in a larger E-field in the neck and shoulders. Coil C lowers the E-field in the body from coil B by limiting the coil turns in the fold region. We should also note here that due to interference of the shoulder of Yoon-sun and the gradient coil, Yoon-sun has to be shifted 4 cm inferior to Duke in these three coils. This is another reason why augmentation of nerves was introduced into both Yoon-sun and Duke models.

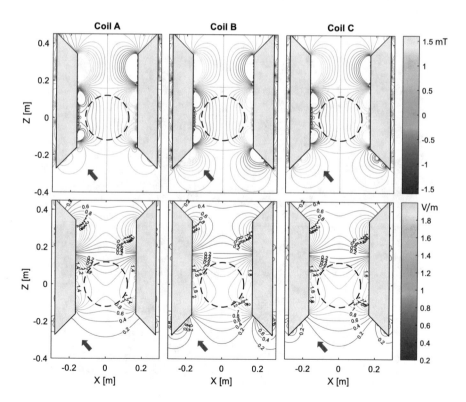

Fig. 7 Field of the folded and non-folded gradient coils in free space. Upper row: Bz in XZ plane (scaled to 10mT/m); lower row: $|\vec{E}|$ in XZ plane (scaled to 100 T/m/s). Red arrows show the different field distributions in three coils

3.3   PNS Simulation Results for Folded/Non-folded X Coils

While Figure 8 provides the E-field intensity, Figure 9 shows the most PNS-sensitive locations in Yoon-sun and Duke in the three newly designed gradient coils. For Coil A, the action potentials are initiated mostly in the forehead region in Yoon-sun. For the same coil, the action potentials are formed in both the forehead and glabellar regions in Duke. For coil B, the strongest PNS is observed in the neck or shoulder region. For Coil C, the most PNS-sensitive regions have a situation between coil A and coil B. Coil C yields the highest threshold so the lowest risk in the Yoon-sun model but has similar performance to coil B for Duke. For coil C, although the E-field in the body is higher than the E-field in the head (Fig. 8), the PNS threshold in the body is not necessarily lower. This is because, from the nerve cable equation, it is the first order difference of E-field intensity along the nerve that determines PNS, not the E-field intensity itself. But overall, PNS-sensitive locations are correlated to the high field region of the gradient coil. The difference in PNS results between the models can be explained by the difference in the size of the subjects and their relative positions in the coil. Figure 5b summarizes the PNS thresholds of the three coils.

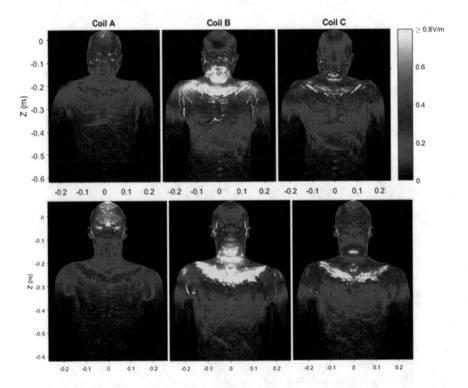

Fig. 8 $|\bar{E}|$ displayed on XZ plane, upper row: Yoon-sun, glabella@-60 mm from gradient coil ISO (due to the coil/shoulder interference); lower row: glabella@-20 mm from gradient coil ISO

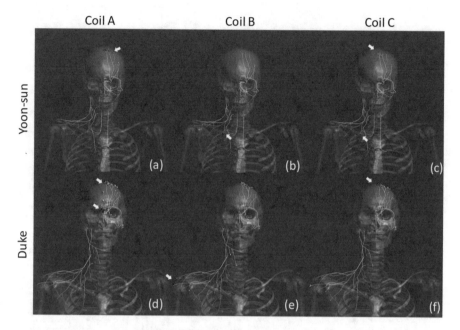

Fig. 9 Locations of PNS initiation in nerve trajectories stimulated by three X coils are represented by small balls. White arrows mark the most sensitive regions in case difficult to read

3.4 E-Field Streamlines

We recommend using E-field streamline plot to further reveal the relationship between the gradient coil wire pattern and PNS as what is shown in Fig. 10. E-field follows the eddy current $\vec{J}$ in the human body since $\vec{J} = \sigma\vec{E}$, so that using E-field streamline is a better way to represent how the eddy current flows in the human body than E-field intensity map. It can be seen that eddy current should follow the wire pattern of the gradient coil and is also affected by the geometry and tissue properties of the human body.

For Coil A, since most of the conductor turns are at the upper head region, higher intensity E-fields and thus eddy currents, will flow through this region. On the other hand, since Coil B has many wire turns in the lower part of the coil, the eddy current loops concentrate on the neck and shoulders. This also leads to reduced eddy currents in the forehead region. Coil C yields a situation in between Coils A and B. It is clear that the E-fields that form the eddy current determine the PNS. Thus, to design the gradient coil with PNS constraints, one would seek to shape the eddy current paths in the human body to avoid producing high E-fields (driving function) in regions where peripheral nerves are located. While previous studies propose algorithms to constrain PNS in the gradient coil design process, here, we provide a more intuitive explanation of how they work.

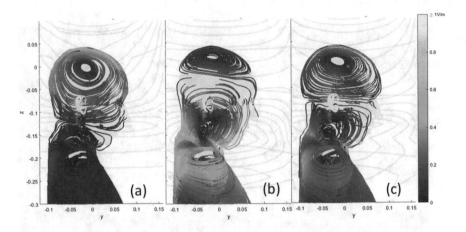

Fig. 10 $\bar{E}$ field streamlines in Yoon-sun model by three new designed coils: (**a**) coil A, (**b**) coil B, (**c**) coil C rendered with overlapping wire pattern of the coils (SR scaled to 100 T/m/s)

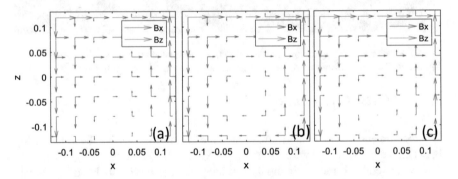

Fig. 11 Concomitant Bx field rendered together with Bz field for three new designed coils

3.5 Concomitant Field

Conventionally, only B_z is controlled in gradient coil design since only B_z is involved in image encoding. Previous studies [18, 19] showed it is possible to reduce the PNS risk by manipulating the concomitant field of the gradient coil. The concomitant field exists because the magnetic field is harmonic at places of no source, and it is impossible to obtain a linearly changing B_z without involving changing B_x and/or B_y [20]. Figure 11 shows the concomitant B_x fields together with the B_z fields for the three X coils. We can see that while B_z fields of these coils are identical in FOV, the B_x fields are different. The correlation between $|\bar{E}|$ and $|\bar{B}|$ can again be roughly explained by Eq. (1) and be verified with Figs. 11 and 12. $|\bar{E}|$ field can be further manipulated by involving an additional uniform field coil [19] since essentially G is used in MR imaging process. Basically, the concomitant field can be changed by the changing of the wire pattern, as the same as that the eddy current loop in the human

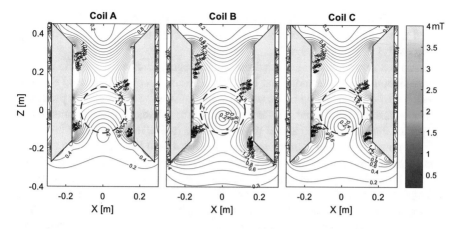

Fig. 12 Free space $|\bar{B}|$ field of three folded and non-folded coils in XZ plane (SR scaled to 10 mT/m). Coil A provides relatively higher $|\bar{B}|$ values in the upper FOV region and Coil B and Coil C provide relatively higher $|\bar{B}|$ values in the lower part (neck and shoulder region)

body can be changed by the wire pattern. But it should be noted that directly controlling the concomitant field does not involve the information of the human body, which includes the geometry and the tissue properties.

3.6 Effects of Homogeneous/Simplified Tissue Properties

The homogeneous human body model is attractive because it may simplify numerical computations and enable the utilization of the boundary element method (BEM), which might be faster than the FEM. But BEM only has a computational advantage when the boundaries or numbers of tissue types are relatively small. In the IEC 60601-2-33 standard, it has been recommended to use a homogeneous cylinder for a rough estimation of the $|\bar{E}|$ field. Although 0.2 S/m is mentioned in this standard, which is roughly the average electrical conductivity in the human body, the E-field will not be affected by any electric conductivity value for the homogeneous model. Another situation is to use the human body model with as few tissue properties as possible. This is interested because segmentations for different tissues are very time expensive. The creation and use of subject-specific human body models for accurate PNS estimation in high-performance gradient systems are gaining more attention, which parallels the need for similar models for local specific absorption rate (SAR) estimation in ultra-high field MRI. The original Yoon-sun and Duke model are composed of over 70 tissues of different electrical conductivities. We conducted several simulation cases to study the impact on PNS results by simplifying the tissues, including (a) a homogenous human body model; (b) a model with only 3 tissue properties (fat, muscle and bone, all the other tissues are treated as muscle); (c) a model with 6 tissue properties (lung, liver and skin are further distinguished according to case (b)); and (d) a model with 8 tissue properties (averaged electric conductivities of tissues inside of the skull and tissues inside of the belly are further added in simulation to represent tissue inside of skull and tissue inside of belly, respectively).

The PNS responses of simplified tissue models are compared with the original heterogeneous Yoon-sun model on MAGNUS X and Z coils. It is noted that it is beneficial to retain accurate tissue model properties for tissues in closest proximity to nerve tissue. Tissues that are far away from the nerves can be assigned tissue properties that are averaged across the local tissue types. This is a compromise that reduces computation time while increasing the likelihood that the $\vec{E}$ field distribution in closest proximity to nerve tissues conform closely to that in a fully segmented human body model. Table 4 lists the conductivities used in the simplified models. To compare the PNS response from different cases, we first convert the PNS response lines of different nerve trajectories in $\tau - \Delta G$ plane to be points in $\Delta G_{min} - SR$ plane. Then we examine how the responses of the nerves shifted in the $\Delta G_{min} - SR$ plane due to the property simplification. We can calculate the pseudo-distance between the corresponding points to quantify the difference but only the most sensitive nerves (for example 20) matter. We can also count how many trajectories remain in the list of the original 20 most sensitive nerve trajectories. As an example (Fig. 13 and Table 5), for the 8-tissue model, stimulated by MAGNUS X coil, the averaged distance is 13.61 for the 20 most sensitive nerves and 20 of them are remained as the most sensitive nerves in Yoon-sun, while these numbers for the homogenous model become 48.62 and 16, respectively. Table 5 summarizes all the cases performed for the Yoon-sun model. It shows that compared to the homogeneous model, the 3-tissue model provides a closer result to the original heterogeneous model and that the 6-tissue model is even closer to the latter. The 8-tissue model is a little better than the 6-tissue one because the averaged electric conductivity for tissues in the skull and in the belly is close to that of the muscle. Another observation is that fewer tissue properties are needed for the MAGNUS X coil than for the Z coil. This is because there are inherently fewer types of tissue outside of the skull than in the body. Overall, the result shows that the homogenous model could be used to give a rough estimation on the E-field but might be not accurate enough to estimate the PNS. Six tissue properties or more might provide a better estimation.

Table 4 Tissue conductivities used in simplified tissue models

Tissue name	σ (S/m)
Fat	0.057
Muscle	0.355
Bone	0.004
Skin	0.170
Lung	0.105
Liver	0.220
Averaged tissue in the skull	0.312
Averaged tissue in the belly	0.310

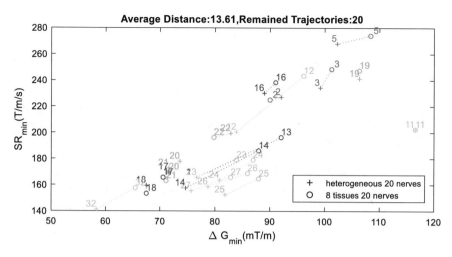

Fig. 13 PNS pseudo-distance of the 20 most sensitive nerve trajectories between simplified tissue model with 8 tissues and the original heterogonous Yoon-sun model. (The number marked on each point is the ID assigned to the nerve trajectory. Different colors are for different nerve branches)

Table 5 The averaged pseudo-distances and top sensitive nerve numbers (of the 20 most sensitive trajectories in the heterogonous model) of simplified tissue models

	Simplified tissue model	Averaged pseudo-distance	Top sensitive nerve no. (of 20)
MAGNUS X	8 tissues	13.61	20
	6 tissues	16.61	19
	3 tissues	32.68	17
	Homogeneous	48.62	16
MAGNUS Z	8 tissues	5.76	19
	6 tissues	5.9	19
	3 tissues	57.16	8
	Homogeneous	110.42	8

4   Shortcomings and Future Works

There are two shortcomings in this study. The first is that only a portion of the nerves has been reconstructed, limiting PNS analysis for X Coils. More nerves are needed to enable full X/Y/Z coil design and analysis. Moreover, the nerve trajectories' geometry might not be perfectly accurate anatomically since a detailed image-derived nerve model was not available. However, compared to the lack of nerve trajectories in the interested places in the original body models, these added trajectories provide richer information to capture PNS responses for gradient coil analysis and design. A similar situation has been reported in [17], where virtual human bodies models are used. The second shortcoming is that the PNS constraint is not yet integrated into the gradient coil design in this study. It is preferred to include this constraint in the future.

5 Conclusions

In this work, extracranial nerves and upper body superficial nerves have been added into both Yoon-sun and Duke models. PNS analysis for gradient coil in the human body models can give results of reasonable accuracy with a relative complete nerve trajectory atlas in the places of interest. Three folded and non-folded head gradient coils of different wire patterns have been designed to reveal the relationship between the E-field/eddy current flow in human body and PNS. E-field streamline plots inside of the human body can provide more information than intensity projection plots. To manipulate PNS through concomitant field generally is to tune the E-field distribution through changing the B-field distribution but does not include the information of the human body. A homogenous model is not accurate enough for PNS threshold estimation, instead, six or more tissues are needed for simplified tissue models.

Acknowledgements This work was funded in part by CDMRP W81XWH-16-2-0054.

References

1. T.C. Cosmus, M. Parizh, Advances in whole-body MRI magnets. IEEE Trans. Appl. Supercond. **21**(3), 2104–2109 (2011). https://doi.org/10.1109/TASC.2010.2084981
2. J. Jin, *Electromagnetic Analysis and Design in Magnetic Resonance Imaging*, 1st edn. (CRC Press, Boca Raton, 1998)
3. R.W. Brown, Y.-C.N. Cheng, E.M. Haacke, M.R. Thompson, R. Venkatesan, *Magnetic Resonance Imaging: Physical Principles and Sequence Design*, 2nd edn. (Wiley-Blackwell, Hoboken, 2014)
4. F. Liu, H. Zhao, S. Crozier, On the induced electric field gradients in the human body for magnetic stimulation by gradient coils in MRI. IEEE Trans. Biomed. Eng. **50**(7), 804–815 (2003). https://doi.org/10.1109/TBME.2003.813538
5. W. Irnich, F. Schmitt, Magnetostimulation in MRI. Magn. Reson. Med. **33**(5), 619–623 (1995). https://doi.org/10.1002/mrm.1910330506
6. S.S. Hidalgo-Tobon, Theory of gradient coil design methods for magnetic resonance imaging. Concepts Magn. Reson. **36A**(4), 223–242 (2010). https://doi.org/10.1002/cmr.a.20163
7. T.K.F. Foo et al., Lightweight, compact, and high-performance 3 T MR system for imaging the brain and extremities. Magn. Reson. Med. **80**(5), 2232–2245 (2018). https://doi.org/10.1002/mrm.27175
8. T.K.F. Foo et al., Highly efficient head-only magnetic field insert gradient coil for achieving simultaneous high gradient amplitude and slew rate at 3.0T (MAGNUS) for brain microstructure imaging. Magn. Reson. Med. **83**(6), 2356–2369 (2020). https://doi.org/10.1002/mrm.28087
9. B.A. Chronik, B.K. Rutt, Simple linear formulation for magnetostimulation specific to MRI gradient coils. Magn. Reson. Med. **45**(5), 916–919 (2001). https://doi.org/10.1002/mrm.1121
10. C.C. McIntyre, A.G. Richardson, W.M. Grill, Modeling the excitability of mammalian nerve fibers: influence of Afterpotentials on the recovery cycle. J. Neurophysiol. **87**(2), 995–1006 (2002). https://doi.org/10.1152/jn.00353.2001
11. D.R. McNeal, Analysis of a model for excitation of myelinated nerve. IEEE Trans. Biomed. Eng. **BME-23**(4), 329–337 (1976). https://doi.org/10.1109/TBME.1976.324593

12. F. Tang et al., An improved asymmetric gradient coil design for high-resolution MRI head imaging. Phys. Med. Biol. **61**(24), 8875–8889 (2016). https://doi.org/10.1088/1361-6560/61/24/8875
13. H.S. Lopez, L. Feng, M. Poole, S. Crozier, Equivalent magnetization current method applied to the design of gradient coils for magnetic resonance imaging. IEEE Trans. Magn. **45**(2), 767–775 (2009). https://doi.org/10.1109/TMAG.2008.2010053
14. Y. Hua, D.T.B. Yeo, T.K. Foo, PNS Estimation of a High Performance Head Gradient Coil by a Coupled Electromagnetic Neurodynamic Simulation Method, in *2020 50th European Microwave Conference (EuMC)*, Jan. 2021, pp. 1071–1074. https://doi.org/10.23919/EuMC48046.2021.9338172
15. M. Davids, B. Guérin, V. Klein, L.L. Wald, Optimization of MRI gradient coils with explicit peripheral nerve stimulation constraints. IEEE Trans. Med. Imaging **40**(1), 129–142 (2021). https://doi.org/10.1109/TMI.2020.3023329
16. E.T. Tan et al., Peripheral nerve stimulation limits of a high amplitude and slew rate magnetic field gradient coil for neuroimaging. Magn. Reson. Med. **83**(1), 352–366 (2020). https://doi.org/10.1002/mrm.27909
17. M. Davids, B. Guérin, A. vom Endt, L.R. Schad, L.L. Wald, Prediction of peripheral nerve stimulation thresholds of MRI gradient coils using coupled electromagnetic and neurodynamic simulations. Magn. Reson. Med. **81**(1), 686–701 (2019). https://doi.org/10.1002/mrm.27382
18. P.B. Roemer, B.K. Rutt, Minimum electric-field gradient coil design: theoretical limits and practical guidelines. Magn. Reson. Med. **86**(1), 569–580 (2021). https://doi.org/10.1002/mrm.28681
19. S.S. Hidalgo-Tobon, M. Bencsik, R. Bowtell, Reducing peripheral nerve stimulation due to gradient switching using an additional uniform field coil. Magn. Reson. Med. **66**(5), 1498–1509 (2011). https://doi.org/10.1002/mrm.22926
20. M.A. Bernstein, K.F. King, Z.J. Zhou, *Handbook of MRI Pulse Sequences* (Academic, Amsterdam/Boston, 2004)

Part III
Low Frequency Electromagnetic Modeling and Experiment: Transcranial Magnetic Stimulation

Experimental Verification of a Computational Real-Time Neuronavigation System for Multichannel Transcranial Magnetic Stimulation

Mohammad Daneshzand, Lucia I. Navarro de Lara, Qinglei Meng, Sergey Makarov, Işıl Uluç, Jyrki Ahveninen, Tommi Raij, and Aapo Nummenmaa

1   Introduction

Transcranial Magnetic Stimulation (TMS) is a non-invasive and safe method for activation of cortical regions by delivering high amplitude, short current pulses into a coil adjacent to the subject's scalp. This creates a strong time-varying magnetic field (~1–2 T), which in turn induces an E-field at the surface of the brain [1, 2]. The E-fields can depolarize/hyperpolarize the cell membrane to activate/inhibit neuronal populations [3]. TMS is approved by the FDA for treating neuropsychiatric disorders such as major depressive disorder (MDD) [4] and obsessive-compulsive disorder (OCD) [5], with more clinical applications under investigation. While accurate targeting of a brain region with a single TMS coil is a highly useful and accurate method for various experimental paradigms, characterising the functional connectivity of adjacent areas of a cortical network is only achieved by dual or multichannel TMS systems [6, 7]. Additionally, multichannel TMS arrays allow electronic steering of the induced Electric field (E-field) without the need to physically move the coils [2, 8] resulting in the capability of rapidly shifting the target/focus. This can be accomplished by computationally determining the E-field distribution of each coil in the array individually and then combining them all together to

M. Daneshzand (✉) · L. I. Navarro de Lara · Q. Meng · I. Uluç · J. Ahveninen
T. Raij
Athinoula A. Martinos Center for Biomedical Imaging, Department of Radiology, Harvard Medical School, Massachusetts General Hospital, Charlestown, MA, USA
e-mail: mdaneshzand@mgh.harvard.edu

S. Makarov
ECE Department, Worcester Polytechnic Institute, Worcester, MA, USA

A. Nummenmaa
Harvard Medical School, Massachusetts General Hospital, Charlestown, MA, USA

© The Author(s) 2023
S. Makarov et al. (eds.), *Brain and Human Body Modelling 2021*,
https://doi.org/10.1007/978-3-031-15451-5_4

synthesize a desired E-field pattern to stimulate any given cortical target according to a user-specified cost-function [2].

In general, a major challenge in TMS targeting is how to position the coil with respect to subject's head for optimal stimulation of a desired brain region [9–11]. To address this, a neuronavigation system can be used in which an optical tracking device localizes the TMS coil with respect to the subject's head [12], ensuring consistent coil placement across multiple TMS sessions [13]. However, using a basic navigation system without anatomically realistic E-field modeling, the actual stimulation intensity at the intracranial target and the surrounding regions remains unknown. Therefore, it is critical to computationally estimate the distribution of the TMS-induced E-field intensity in order to delineate the affected brain regions in real-time [12]. In general, we need to consider several factors for precise computational TMS navigation. First, the anatomy of subject's head must be accurately registered to their MRI data, ensuring that systematic coil positioning errors are minimized [11]. Second, we need to rapidly and accurately calculate the E-field pattern of the TMS coil or coil array to create the best possible "match" with the target area [12]. The real-time E-field based computational neuronavigation is utilized in our multichannel TMS system within a slightly different way such that when the coil array position is fixed with respect to the subject's head, the E-fields need to be computed only once and different current amplitudes will be applied to each coil to synthesize a desired E-field pattern at the target [8]. However, if the subject's head moves between pulses, the currents need to be adjusted to compensate for this, requiring a "near real-time recalculation" of the induced E-fields for all the coils in the array.

For interactive E-field based navigation of the TMS coil during a stimulation session, the E-field distributions need to be ideally computed and displayed within 100 ms (frame rate of 10 Hz). While simple (spherical) models used in commercial neuronavigation systems allow fast E-field computations, they may give inaccurate results due to oversimplification of the volume conductor (head) model [14, 15]. On the other hand, while high resolution individualized head models used in Boundary Element Method (BEM) and Finite Element Method (FEM) reach higher spatial precision, they are computationally too sluggish for real-time navigation requiring ~10 seconds per solution [16–18]. We have previously addressed this speed vs. precision dilemma by introducing a new method based on the concept of an individualized Magnetic Stimulation Profile (MSP) [19]. The real-time E-field estimation by MSP approach leverages pre-calculation of the E-fields from a set of dipoles placed around the head model. The pre-calculations are done with the Boundary Element Method accelerated by the Fast Multipole Method (BEM-FMM) [20, 21]. Since the total E-field of an arbitrary TMS coil only depends on the incident E-field and tissue conductivity boundaries [20], we can find the matching coefficients between the incident E-field of the coil and the dipole basis set approximation, and the total E-field of the coil will be obtained by applying these matching coefficients to the total E-fields of the dipoles [19].

Here we used two 3-axis coils [8] to illustrate how different combinations of the array elements allow steering the E-field 'hot spot'. Furthermore, our computational multichannel TMS neuronavigation system allows rendering the E-fields of individual 3-axis coils and synthesizing the 'hot spot' using the array approach with speed and accuracy suitable for human studies. To verify that the computational neuronavigation system is working as expected, we used two z-elements of our 2×3-axis coil array connected to individual stimulators to mimic a figure of eight coil to activate the motor cortex and compared the results with a commercially available TMS coil.

This article is organized as follows: In Sect. 2 we describe the design of the 3-axis coils and investigate the efficiency of these coils in a simple 2×3-axis array configuration by calculating the induced E-field using a spherical model. Section 3 provides a brief description of TMS neuronavigation methods as well as their present limitations in practical applications. In Sect. 4 we briefly describe the methodology behind the MSP approach for real-time calculation of the E-fields and its integration with a commercial TMS navigation system. Finally, in Sect. 5, we demonstrate the efficiency of two z-elements in a 2×3-axis array in conjunction with the interactive navigation system to elicit Electromyography (EMG) responses in a healthy volunteer.

2   Multichannel TMS Array Design Concept

2.1   The 3-Axis Coils as the Building Blocks of a TMS Array

The proposed modular 3-axis multichannel array allows efficient and safe stimulation of any cortical area with high degrees of freedom in shaping and adjusting the orientation of the E-field. Delivering specific current amplitudes for each coil enables simultaneous or sequential stimulation of multiple areas which can be utilized for investigation of causal relationships between cortical areas involved in various information processing tasks [22, 23].

Figure 1a shows the basic conceptual design of our envisioned 48-channel (16 × 3) TMS array. In this array configuration that has been motivated by our overarching goal of an integrated TMS-compatible MRI acquisition system [24], there are 16 locations for placement of 3-axis coils allowing whole-head coverage. The 3-axis coil consists of three orthogonal circular elements called X, Y and Z as shown in Fig. 1b, c. The goal of this coil design is to achieve high efficiency and focality while offering accurate spatial control of the E-field hot spots using simple coil modules. The purpose of adding the X and Y elements is to cover for the zero E-field at the center of the circular Z-element [8]. Additionally, the X and Y elements will provide more degrees of freedom to the overall field shaping capabilities of the multichannel array. Furthermore, the geometrical design of the X, Y and Z elements eliminates the coupling between them within each unit [8].

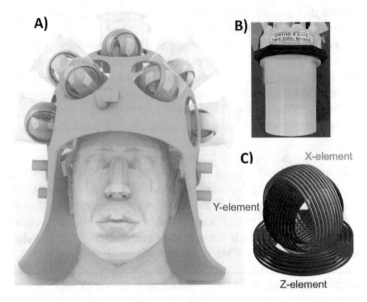

Fig. 1 (**a**) Multi-channel TMS coil array. (**b**) prototype of the first of its kind 3-axis coil. (**c**) Each 3-axis coil consists of three orthogonal elements allowing separate and combined stimulations

2.2 The Numerical Method for E-Field Computation of TMS Coils

The calculation of the E-field induced by each coil element was done using the BEM approach accelerated by the Fast Multipole Method (BEM-FMM) [20]. Based on the Maxwell-Faraday law, the magnetic field of the TMS coil induces an electric field:

$$\nabla \times \boldsymbol{E} = -\frac{\partial \boldsymbol{B}}{\partial t}, \tag{1}$$

where B is the magnetic field of the coil. The total electric field E can be written as:

$$\boldsymbol{E} = \boldsymbol{E}^{inc} + \boldsymbol{E}^{s} = -\frac{\partial \boldsymbol{A}}{\partial t} - \nabla \varphi \tag{2}$$

The $\boldsymbol{E}^{inc} = -\partial A/\partial t$ corresponds to the primary or 'incident' field induced by the current in the coil, and the $\boldsymbol{E}^{s} = -\nabla \varphi$ is called the secondary field that arises from the differences in the electrical conductivities of various tissues inside the head [25]. The BEM-FMM operates with tissue conductivity boundaries where $\boldsymbol{E}^{inc}$ causes accumulation of surface charges giving rise to the secondary field $-\nabla \varphi$. The boundary integral equations stemming from the quasi-static current conservation conditions $\nabla \cdot \boldsymbol{J} = \boldsymbol{0}$ are numerically solved by the Generalized Minimal Residual Method (GMRES) [26]. Once the solution converges and the charge distribution on the conductivity boundaries is known, the total E-field in any region of the 3D space can be obtained.

The individual E-field patterns generated by each element were previously characterized [8] and here we explore different E-field patterns obtained by various combinations of the elements. The coil models were generated in MATLAB [21] (MathWorks, Inc., Natick, MA, USA), placed over a 1-layer spherical model and the E-fields were calculated on the inner sphere at approximately 2 cm distance from the center of the coils. Figure 2 shows the model configuration as well as several calculated E-field patterns based on the activated elements and current polarities. The coil current rate of change $\left(\dfrac{dI}{dt} \right)$ values were corresponding to 50% of Maximum Stimulator Output (MSO). However, for actual stimulation experiments, the current intensity delivered to each element can be optimized using Minimum Norm Estimate (MNE) method [27] in order to generate a desired E-field pattern.

Figure 2 also shows the E-field pattern of a commercially available figure-of-eight coil (C-B60, MagVenture, Farum, Denmark) as a reference. A similar but somewhat more focal pattern can be generated by combining two of the z-elements (Fig. 2, b8). However, the C-B60 generates a fixed E-field pattern with maximum intensity at the intersection of the two wings and to stimulate a different region the coil must be physically moved. Generally, even a simple multichannel array

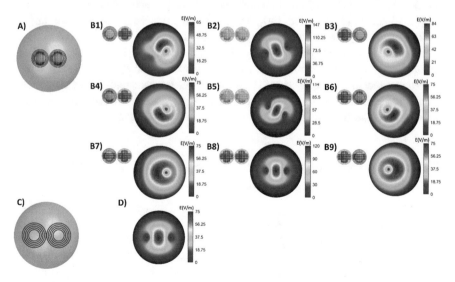

Fig. 2 (**a**) Two 3-axis coils positioned on a spherical head model. (**b1–9**) Corresponding E-field distributions of the two 3-axis coils with the activated elements shown in green. (**c**) A commercially available standard figure-of-eight TMS coil model (MagVenture C-B60) and the corresponding E-field pattern (**d**)

comprising of two 3-axis coils provides high flexibility in E-field shaping while still offering stimulation intensities comparable to commercially available figure-of-eight coils.

3   TMS Navigation Systems

The rationale of using an image-guided TMS navigation system is to reduce the variability of TMS coil placement with respect to the subject's head and hence to improve the repeatability and accuracy of TMS targeting and dosing [28, 29]. TMS navigator systems typically use optical tracking devices to localize the coil with respect to the subject's head. Anatomical landmarks on the subject's head are identified with corresponding points in the Magnetic Resonance Imaging (MRI) data of the subject, allowing co-registration of the TMS coil position/orientation with the individual anatomy.

Here, we used a commercial TMS neuronavigation system (LOCALITE GmbH, Germany) that employs an optical tracking camera (Polaris Spectra, Northern Digital, Inc. Waterloo, Ontario) to capture the position of the reference and coil trackers and displays the information graphically within the neuronavigation software. Additionally, a specific stimulation target can be defined using the software. The interactive navigation tool allows for accurate positioning of the coil at the target during the stimulation experiment. The positioning of the coil with the neuronavigation system is naturally prone to some errors such as the method used for optical tracking of the coils and the head as well as the registration accuracy of the subject's head to the MR data [30]. Moreover, quantitative high-resolution targeting/dosing of TMS requires knowing the E-field intensity at the desired cortical location and surrounding regions [31] which is necessary especially for multichannel applications but not available on commercial navigation systems. Here, we leveraged our recently developed MSP approach [19] for near real-time TMS E-field calculation combined with a commercial navigation system to track the coil movements and render the induced E-field with a frame rate of 6 Hz.

4   Fast E-Field Calculation with Dipole Based Magnetic Stimulation Profile Approach

A detailed description of the MSP approach can be found in [19]. The computational pipeline of the MSP approach consists of a pre-calculation and a real-time step. In the pre-calculation step, hundreds of magnetic dipoles are distributed uniformly on the scalp model as shown in Fig. 3a, and the incident (field from the dipoles only) and total E-fields (fields from dipoles coil and charge accumulation at tissue conductivity boundaries) are computed using the BEM-FMM method [20]. This fundamental basis solution only needs to be calculated once per subject and can be subsequently used to quickly estimate the E-field created by any TMS coil.

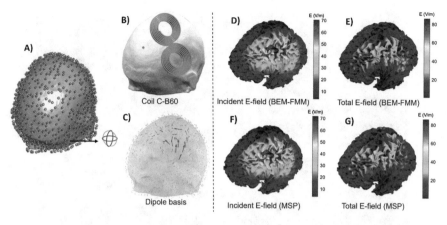

Fig. 3 (**a**) Spatial configuration of an example basis set with 3 orthogonal magnetic dipoles positioned at several hundreds of locations around the subject's scalp surface for pre-calculation of the incident and total E-fields. (**b**) A model of a commercially available TMS coil (C-B60, MagVenture, Farum, Denmark) is placed at an arbitrary location and the dipole amplitudes (**c**) are optimized to match the incident E-field of the coil. (**d, e**) the incident and total E-field of the C-B60 coil calculated by the BEM-FMM as the ground truth. (**f**) The incident E-field obtained from the dipoles basis set. (**g**) The matching coefficients obtained in (**c**), are applied to the dipole basis set total E-fields to approximate the total E-field of the coil

The real-time step is based on fundamental physical principles of TMS-induced E-fields:

- The total E-field of a TMS coil only depends on the incident field and the tissue conductivity boundary surfaces and the associated conductivities [1].
- The incident E-field created by the coil in free space can be re-computed very quickly by applying translations/rotations to the set of points where the Incident E-field was initially calculated on.
- By projecting the incident field of the TMS coil onto the set of dipoles and obtaining a set of weights that provide the optimal 'match', the same weights can be used to approximate the total E-field based on the principle of superposition [32, 33].

Based on these principles, the pre-calculated E-fields of the dipole basis set can be utilized for a very fast estimation of the total E-field of an arbitrary TMS coil as shown in Fig. 3. This method allows for real-time (100 ms) computational performance and display of the TMS-induced E-field for interactive visualization of the stimulated cortical regions.

Figure 4 shows an example of the real-time setup consisting of a MSP-based computational E-field engine integrated with a commercial navigator system (LOCALITE, Bonn, Germany), using a head phantom. The reference head tracker placed on the phantom allows for co-registration of the head shape with a 'synthesized MRI dataset'. The position of the coil with respect to phantom is recorded and displayed by the navigator software using the tracker attached to the coil and an optical position measurement camera. The position of the coil(s) (either the two

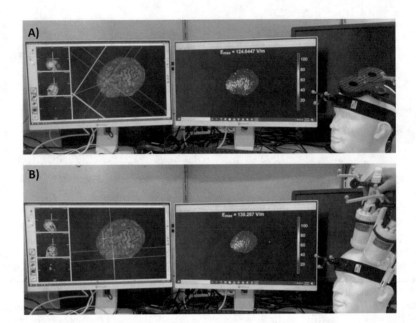

Fig. 4 The real-time E-field computational modeling setup consisting of a commercial neuronavigation system (LOCALITE, Bonn, Germany) interfaced with an external PC for MSP-based calculation and display of (**a**) a commercial figure-of-eight coil (C-B60) and (**b**) two custom-made 3-axis coils (Tristan Technologies, San Diego, California)

z-elements of the two 3-axis coils or the C-B60 coil) are then streamed into a second PC using the TCP/IP and JSON protocols implemented in the LOCALITE navigator. Finally, with the coil position/orientation streamed continuously to MATLAB, we can calculate the total E-field and display the results on a cortical surface as shown in Fig. 4. In the current setup, we were able to freely move the coil around the phantom and calculate/visualize the total E-field within a frame rate of 3 Hz and 6 Hz for the two z-elements of the two 3-axis coils and the C-B60 coil, respectively, using MATLAB 2021a on an Intel Xeon(R) Gold 6226R PC with 192 GB of memory.

5 Motor Cortex Stimulation Experiment with Z-Elements of Two 3-Axis Coils

We recruited one healthy male volunteer to evaluate the performance of the simultaneous activation of currents of opposite polarity in the Z-elements of two 3-axis coils in a motor cortex stimulation experiment. Informed consent was obtained from the participant in accordance with the study protocol that was approved by the local IRB at Massachusetts General Hospital. Initially, we acquired the subject's T1-weighted MRI data (1 mm isotropic) on a Siemens Trio 3 T scanner and reconstructed the tissue compartment boundary surface meshes using the SimNIBS, and Freesurfer tools [34, 35]. The subject's anatomical landmarks were registered to the MRI data using the Localite neuronavigation system and two separate trackers were mounted on the 3-axis coils to track their position with respect to subject's head. We set up a TCP/IP protocol to communicate the coordinates of these coils into MATLAB and calculated the combinatory total E-field of the two 3-axis coils. A similar setup was used for the commercial C-B60 TMS coil that was used as a reference.

The EMG data was continuously recorded from the First Dorsolateral interosseous (FDI) muscle of the right hand using the BrainAmp ExG amplifier (Brain Products, Gilching, Germany) and the Motor Evoked Potentials (MEPs) corresponding to each TMS pulse were saved for post-hoc analysis. Either one or two commercially available TMS stimulators were used (MagPro X100, MagVenture) to deliver the current pulses. To identify the FDI MEP hotspot, we started with the stimulation intensity set to 50% MSO and moved the C-B60 TMS coil over the M1 cortex guided with the neuronavigation system with the concurrent display of the E-field similar to the setup described in Fig. 4. Single pulses were delivered, and the intensity was increased until an MEP amplitude of ~50–100 μV was observed. The smallest stimulator output intensity by which the EMG response of at least 50 μV with 50% probability is observed, was then defined as the FDI resting motor threshold (MT) [36]. For quantitative reference, we placed the C-B60 coil at the previously defined FDI 'hot spot' and recorded the MT as well as several clearly suprathreshold MEPs (at 107% resting MT). Next, the 2×3-axis coil array was positioned over the FDI hot spot and the z-elements of the two 3-axis coils were synchronously activated to mimic a figure-of-eight coil configuration. The stimulation intensity was increased until clearly suprathreshold MEPs were observed. The post-hoc analysis of the MEP responses in Fig. 5 shows that both dual z-elements and the

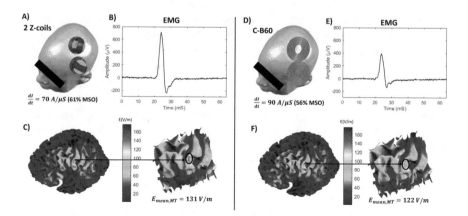

Fig. 5 (**a**) The z-elements of two 3-axis coils stimulating the left M1 hand area. (**b**) The recorded MEPs from FDI muscle. (**c**) The E-field distribution on the white matter surface. (**d, e**) Corresponding results for the C-B60 coil

C-B60 coil elicited suprathreshold responses with similar EMG morphologies recorded at the FDI muscle. Furthermore, the computationally estimated cortical E-fields show robust agreement for both the amplitudes and the spatial patterns.

6 Summary and Discussion

A key requirement of multi-channel TMS technology is fast and accurate computation of the E-fields from all coil elements to determine the optimal linear combination of the input current values in order to stimulate a desired cortical target region. In this study, we demonstrated that precise and fast computation of the E-field can be achieved by (1) reconstructing high resolution surfaces meshes based on the individual MRI data, (2) reducing the computational burden of high resolution E-field modeling by leveraging the MSP-based approach utilizing a spatially fixed dipole basis set, and (3) retrieving the pre-calculated solutions for the near-real time step to obtain the total E-field by means of simple matrix multiplications based on matching the incident field of the TMS coil with the dipole basis set [19]. The position of the TMS coils with respect to subject's head can be acquired with an optical tracking camera and displayed in a neuronavigation software with frame rates of 10–15 Hz. This gives us a time frame of 60–100 ms for E-field calculations to be performed for an interactive navigation pipeline. Our results suggest that we are able to calculate the E-field of a C-B60 coil within this timeframe on a high-resolution WM surface with about 100 K triangular mesh elements. For the multi-channel TMS arrays the E-field computation time will increase based on the number of coils and coil elements used. However, we plan to improve the computational efficiency further by utilizing a parallel computation approach to obtain the E-field of each coil in the array.

Additionally, we showed that the z-elements of the two 3-axis coils are capable of clearly suprathreshold stimulation when used in a 'dual-channel array' driven by synchronized triggering of two independent TMS stimulators. For more flexible stimulation, the orthogonal placement of the elements in the 3-axis coil provides efficient decoupling between the elements while granting three degrees of freedom in shaping the induced E-field pattern. Moreover, arranging several 3-axis coils in an array will further increase the degrees of freedom for efficient stimulation of a target area with a desired field orientation without the need for physical movement of the coils. We have recently developed a novel 9-channel stimulator system in partnership by MagVenture, by which each element of the 3-axis coils can be driven by independent stimulator units, allowing delivery of pulses with high combined power.

The experimental results show that the juxtaposition of two Z-elements provides an E-field intensity and spatial distribution similar to the commercially available C-B60 coil, producing clear suprathreshold stimulation as measured with MEPs. Furthermore, the MSP-approach used in multi-channel TMS array can be used as a pre-planning step to rapidly calculate the E-field at several locations based on specific criteria of the experiment [37] or to optimize the positions/orientations of the individual 3-axis coil elements. The biophysical signals recorded during the experiment can also be paired with the previously calculated E-field patterns for post-hoc analysis [38, 39], or be utilized as feedbacks to adjust the stimulation parameters (*e.g.*, input current to each coil) in a closed-loop stimulation paradigm [40–42].

Acknowledgements The research was supported by NIH R01MH111829, P41EB030006, and R01DC016915.

References

1. B.J. Roth, L.G. Cohen, M. Hallett, W. Friauf, P.J. Basser, A theoretical calculation of the electric field induced by magnetic stimulation of a peripheral nerve. Muscle Nerve **13**(8), 734–741 (1990). https://doi.org/10.1002/mus.880130812
2. J. Ruohonen, R.J. Ilmoniemi, Focusing and targeting of magnetic brain stimulation using multiple coils. Med. Biol. Eng. Comput. **36**(3), 297–301 (1998)
3. F. Rattay, The basic mechanism for the electrical stimulation of the nervous system. Neuroscience **89**(2), 335–346 (1999)
4. J.P. O'Reardon et al., Efficacy and safety of transcranial magnetic stimulation in the acute treatment of major depression: a multisite randomized controlled trial. Biol. Psychiatry **62**(11), 1208–1216 (2007). https://doi.org/10.1016/j.biopsych.2007.01.018
5. L. Carmi et al., Efficacy and safety of deep transcranial magnetic stimulation for obsessive-compulsive disorder: a prospective multicenter randomized double-blind placebo-controlled trial. Am. J. Psychiatry **176**(11), 931–938 (2019). https://doi.org/10.1176/appi.ajp.2019.18101180
6. S. Groppa, N. Werner-Petroll, A. Münchau, G. Deuschl, M.F.S. Ruschworth, H.R. Siebner, A novel dual-site transcranial magnetic stimulation paradigm to probe fast facilitatory inputs from ipsilateral dorsal premotor cortex to primary motor cortex. NeuroImage **62**(1), 500–509 (2012). https://doi.org/10.1016/j.neuroimage.2012.05.023
7. Y. Roth, Y. Levkovitz, G.S. Pell, M. Ankry, A. Zangen, Safety and characterization of a novel multi-channel TMS stimulator. Brain Stimul. **7**(2), 194–205 (2014). https://doi.org/10.1016/j.brs.2013.09.004
8. L.I.N. de Lara et al., A 3-axis coil design for multichannel TMS arrays. NeuroImage **224**, 117355 (2020)
9. L.D. Gugino et al., Transcranial magnetic stimulation coregistered with MRI: a comparison of a guided versus blind stimulation technique and its effect on evoked compound muscle action potentials. Clin. Neurophysiol. **112**(10), 1781–1792 (2001)
10. T. Picht, J. Schulz, M. Hanna, S. Schmidt, O. Suess, P. Vajkoczy, Assessment of the influence of navigated transcranial magnetic stimulation on surgical planning for tumors in or near the motor cortex. Neurosurgery **70**(5), 1248–1257 (2012)
11. R. Sparing, D. Buelte, I.G. Meister, T. Pauš, G.R. Fink, Transcranial magnetic stimulation and the challenge of coil placement: a comparison of conventional and stereotaxic neuronavigational strategies. Hum. Brain Mapp. **29**(1), 82–96 (2008)
12. J. Ruohonen, J. Karhu, Navigated transcranial magnetic stimulation. Neurophysiol. Clin. Neurophysiol. **40**(1), 7–17 (2010)
13. U. Herwig et al., The navigation of transcranial magnetic stimulation. Psychiatry Res. Neuroimaging **108**(2), 123–131 (2001)

14. L. Heller, D.B. van Hulsteyn, Brain stimulation using electromagnetic sources: theoretical aspects. Biophys. J. **63**(1), 129–138 (1992)

15. A. Nummenmaa, M. Stenroos, R.J. Ilmoniemi, Y.C. Okada, M.S. Hämäläinen, T. Raij, Comparison of spherical and realistically shaped boundary element head models for transcranial magnetic stimulation navigation. Clin. Neurophysiol. Off. J. Int. Fed. Clin. Neurophysiol. **124**(10), 1995–2007 (2013). https://doi.org/10.1016/j.clinph.2013.04.019

16. F.S. Salinas, J.L. Lancaster, P.T. Fox, 3D modeling of the total electric field induced by transcranial magnetic stimulation using the boundary element method. Phys. Med. Biol. **54**(12), 3631 (2009)

17. A. Opitz, M. Windhoff, R.M. Heidemann, R. Turner, A. Thielscher, How the brain tissue shapes the electric field induced by transcranial magnetic stimulation. NeuroImage **58**(3), 849–859 (2011)

18. I. Laakso, A. Hirata, Fast multigrid-based computation of the induced electric field for transcranial magnetic stimulation. Phys. Med. Biol. **57**(23), 7753–7765 (2012). https://doi.org/10.1088/0031-9155/57/23/7753

19. M. Daneshzand, S. N. Makarov, L. I. Navarro de Lara, B. Guerin, B. McNab, B. R. Rosen, M. S. Hämäläinen, T. Raij, and A. Nummenmaa, Rapid computation of TMS-induced E-fields using a dipole-based magnetic stimulation profile approach. NeuroImage **237**(2021)

20. S.N. Makarov, G.M. Noetscher, T. Raij, A. Nummenmaa, A quasi-static boundary element approach with fast multipole acceleration for high-resolution bioelectromagnetic models. I.E.E.E. Trans. Biomed. Eng. **65**(12), 2675–2683 (2018)

21. S.N. Makarov, W.A. Wartman, M. Daneshzand, K. Fujimoto, T. Raij, A. Nummenmaa, A software toolkit for TMS electric-field modeling with boundary element fast multipole method: an efficient MATLAB implementation. J. Neural Eng. (2020). https://doi.org/10.1088/1741-2552/ab85b3

22. K. Davranche, C. Tandonnet, B. Burle, C. Meynier, F. Vidal, T. Hasbroucq, The dual nature of time preparation: neural activation and suppression revealed by transcranial magnetic stimulation of the motor cortex. Eur. J. Neurosci. **25**(12), 3766–3774 (2007)

23. F. Ferreri, P.M. Rossini, TMS and TMS-EEG techniques in the study of the excitability, connectivity, and plasticity of the human motor cortex. Rev. Neurosci. **24**(4), 431–442 (2013)

24. L.I. de Lara, L. Golestanirad, S.N. Makarov, J.P. Stockmann, L.L. Wald, A. Nummenmaa, Evaluation of RF interactions between a 3T birdcage transmit coil and transcranial magnetic stimulation coils using a realistically shaped head phantom. Magn. Reson. Med. **84**(2), 1061–1075 (2020)

25. P.C. Miranda, L. Correia, R. Salvador, P.J. Basser, Tissue heterogeneity as a mechanism for localized neural stimulation by applied electric fields. Phys. Med. Biol. **52**(18), 5603–5617 (2007). https://doi.org/10.1088/0031-9155/52/18/009

26. Y. Saad, M.H. Schultz, GMRES: A generalized minimal residual algorithm for solving nonsymmetric linear systems. SIAM J. Sci. Stat. Comput. **7**(3), 856–869 (1986)

27. M.S. Hämäläinen, R.J. Ilmoniemi, Interpreting magnetic fields of the brain: minimum norm estimates. Med. Biol. Eng. Comput. **32**(1), 35–42 (1994). https://doi.org/10.1007/BF02512476

28. A.T. Sack, R.C. Kadosh, T. Schuhmann, M. Moerel, V. Walsh, R. Goebel, Optimizing functional accuracy of TMS in cognitive studies: a comparison of methods. J. Cogn. Neurosci. **21**(2), 207–221 (2009)

29. B. Langguth, T. Kleinjung, M. Landgrebe, D. De Ridder, G. Hajak, rTMS for the treatment of tinnitus: the role of neuronavigation for coil positioning. Neurophysiol. Clin. Neurophysiol. **40**(1), 45–58 (2010)

30. J. Ruohonen, J. Karhu, Navigated transcranial magnetic stimulation. Neurophysiol. Clin. **40**(1), 7–17 (2010). https://doi.org/10.1016/j.neucli.2010.01.006

31. N. Sollmann et al., Comparison between electric-field-navigated and line-navigated TMS for cortical motor mapping in patients with brain tumors. Acta Neurochir. **158**(12), 2277–2289 (2016)

32. B.B. Baker, E.T. Copson, *The Mathematical Theory of Huygens' Principle*, vol 329 (AMS Chelsea Publishing, Providence, Rhode Island, 2003)
33. L.M. Koponen, J.O. Nieminen, R.J. Ilmoniemi, Minimum-energy coils for transcranial magnetic stimulation: application to focal stimulation. Brain Stimul. **8**(1), 124–134 (2015). https://doi.org/10.1016/j.brs.2014.10.002
34. A. Thielscher, A. Antunes, G.B. Saturnino, Field modeling for transcranial magnetic stimulation: a useful tool to understand the physiological effects of TMS?, in *2015 37th annual international conference of the IEEE engineering in medicine and biology society (EMBC)*, 2015, pp. 222–225
35. B. Fischl, FreeSurfer. NeuroImage **62**(2), 774–781 (2012)
36. S. Rossi, M. Hallett, P.M. Rossini, A. Pascual-Leone, Safety, ethical considerations, and application guidelines for the use of transcranial magnetic stimulation in clinical practice and research. Clin. Neurophysiol. Off. J. Int. Fed. Clin. Neurophysiol. **120**(12), 2008–2039 (2009). https://doi.org/10.1016/j.clinph.2009.08.016
37. A. Nummenmaa et al., Targeting of white matter tracts with transcranial magnetic stimulation. Brain Stimul. **7**(1), 80–84 (2014)
38. J. Ahveninen et al., Evidence for distinct human auditory cortex regions for sound location versus identity processing. Nat. Commun. **4**(1), 1–8 (2013)
39. T. Raij et al., Prefrontal cortex stimulation enhances fear extinction memory in humans. Biol. Psychiatry **84**(2), 129–137 (2018)
40. D. Kraus et al., Brain state-dependent transcranial magnetic closed-loop stimulation controlled by sensorimotor desynchronization induces robust increase of corticospinal excitability. Brain Stimul. **9**(3), 415–424 (2016)
41. J. Meincke, M. Hewitt, G. Batsikadze, D. Liebetanz, Automated TMS hotspot-hunting using a closed loop threshold-based algorithm. NeuroImage **124**, 509–517 (2016)
42. T.O. Bergmann et al., EEG-guided transcranial magnetic stimulation reveals rapid shifts in motor cortical excitability during the human sleep slow oscillation. J. Neurosci. **32**(1), 243–253 (2012)

Evaluation and Comparison of Simulated Electric Field Differences Using Three Image Segmentation Methods for TMS

Tayeb Zaidi and Kyoko Fujimoto

1 Introduction

Magnetic and electrical brain stimulation therapies are widely used to treat neuro-degenerative disorders. One of the commonly used non-invasive techniques is transcranial magnetic stimulation (TMS) that employs magnetic induction to stimulate the brain to improve symptoms for diseases such as depression. Computational modeling has been used to assess the effectiveness and safety of TMS. A detailed brain model is available to allow for these assessments [5]; however, the model is only based on one subject. In order to allow for careful planning of a given treatment regimen, modeling needs to be completed on a per-patient basis.

A patient-specific brain model can be created using medical imaging data. Magnetic Resonance Imaging (MRI) structural data is often used for such purpose. There are a variety of semi-automatic segmentation methods available [1, 2, 4, 22] that can generate a 3D head model using a set of T_1- and T_2-weighted images. Segmentation varies across different methods [6, 18]. Therefore, it may affect electric field distributions in electric-field modeling. Some of the segmentation methods resulted in differences in electric field distributions of up to 30% when evaluated with one computational modeling method [11]. More investigations are needed to confirm the degree of differences among different image segmentation and computational modeling methods.

In this study, the T_1- and T_2-weighted images of 16 subjects were processed with three different segmentation methods. Computational modeling of TMS was performed based on each segmented data by targeting both the primary motor cortex and the dorsolateral left prefrontal cortex (DLPFC), then the simulated electric field results were compared and evaluated.

T. Zaidi · K. Fujimoto (✉)
Center for Devices and Radiological Health, US Food and Drug Administration,
Silver Spring, MD, USA

© The Author(s) 2023
S. Makarov et al. (eds.), *Brain and Human Body Modelling 2021*,
https://doi.org/10.1007/978-3-031-15451-5_5

2   Materials and Methods

2.1   MRI Data and Segmentation

MRI T_1- and T_2-weighted images were used from 16 Human Connectome Project healthy subjects [19] with an isotropic resolution of 0.7 mm per voxel. Two pipelines implemented in the SimNIBS software package v3.2 [15] were used for segmentation, *headreco* [10] and *mri2mesh* [20], as well as a highresolution FreeSurfer [2] pipeline (*fshires*) [21]. Both the *headreco* and *mri2mesh* segmentation methods generate surface and volume segmentation of brain and head structures including gray matter (GM), white matter (WM), cerebrospinal fluid (CSF), skull, and skin. The *fshires* pipeline produces GM and WM segmentation based on the native sub-millimeter resolution.

The surface meshes were generated using the default options from SimNIBs for both *headreco* and *mri2mesh* and the high resolution option within Freesurfer (*fshires*) using the *-hires* flag. The default surface resolution yields surface meshes containing a combined total of roughly 800,000 to 1 million facets.

2.2   Electromagnetic Simulation

A boundary element fast multipole method (BEM-FMM) solver was used for electromagnetic modeling [7–9]. The solver utilizes the generated surface meshes for field estimation. A figure-eight TMS coil was modeled with a diameter of 90 mm for each loop. The coil model was modeled based on a commercial coil (MRiB91 of MagVenture, Denmark). The coil was placed to target both the patient's left primary motor cortex (the hand knob) and the DLPFC via a projection approach and sulcus-aligned mapping [3, 12]. These regions were chosen because they are common targets for TMS therapy. An example positioning of the coil is shown in Fig. 1. Two target points were used for each subject, located within the primary motor cortex and the DLPFC, respectively. The coil position was determined using three steps. First, the coil was placed so that the centerline (shown as the black line in Fig. 1) passed through the given target point on the gray matter interface. Second, the coil centerline was made to be perpendicular to the skin surface. Lastly, the coil position was adjusted so that the dominant field direction was roughly perpendicular to the nearest sulci [7].

2.3   Analysis

Average surface displacement between mesh surfaces generated based on the *headreco* and *mri2mesh* segmentation methods were compared across the 16 subjects for the gray matter, white matter and CSF surfaces. The displacement was calculated by

Fig. 1 An example placement of the TMS coil targeting the subject's left motor cortex is shown. The black line (the coil axis) was used to confirm the coil placement and runs perpendicular to the coil

TMS Coil with Gray Matter

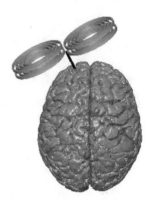

taking the mean of the shortest distance from every triangle centroid of the relevant surface from one segmentation method to all triangle centroids of the surface from the other segmentation method. The electric field values were compared by extracting the values in a 100 mm line perpendicular to the TMS coil axis, along the black line shown in Fig. 1. Comparisons were performed in pairs between *mri2mesh* and *headreco* and between *mri2mesh* and *fshires*. These values were extracted for both of the target points for each subject. The average electric field difference, maximum absolute difference, and maximum percentage difference were compared for all 16 subjects. The average field differences at the target point for both the motor cortex and DLPFC were tested for statistical significance using a paired t-test.

Finally, for additional visualization of the electric field stimulation mapping, all subject electric field results were mapped onto the inflated surface from the FreeSurfer common space (*fsaverage*). The electric field was exported and mapped onto this common surface to map average electric field and its difference surface over the 16 subjects. These maps allow for the qualitative evaluation of focality variations between three different segmentation methods over the entire subject space.

3 Results

All extracerebral and cortical surfaces were successfully reconstructed using the three segmentation methods. The average surface displacement was only calculated for regions of the brain located in the superior cerebral cortex for all 16 subjects between *headreco* and *mri2mesh* and the results are summarized in Fig. 2. The displacement between *fshires* and *mri2mesh* were not compared as the surfaces use the same algorithm from FreeSurfer and the comparison was done by other study [21]. The CSF surface displacement was three times more than the white matter surface displacement, with an average difference of 0.9 mm ($\pm$0.2 mm) for the CSF and 0.3 mm ($\pm$0.06 mm) for the white matter. An example of all the surfaces overlaid on a subject T_1-weighted image is shown in Fig. 3.

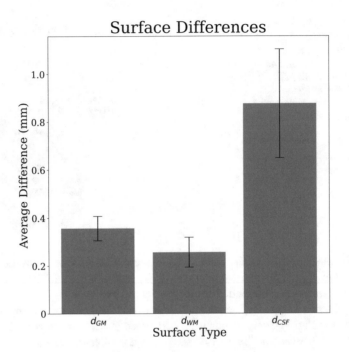

Fig. 2 Average displacement between *headreco* and *mri2mesh* surfaces in millimeters across all 16 subjects for CSF, GM, and WM

For the target point located in the motor cortex, the electric fields showed similar distributions for the *mri2mesh* and the *headreco* segmentation methods. There was an average difference in magnitude of 0.8 V/m and a maximum difference of 61 V/m. The average percentage difference was 2%. In the region of interest within 5 mm of the target point, the average percentage difference was approximately 5%. For the *mri2mesh* and *fshires* segmentation methods, the average percentage difference in the region of interest surrounding the target point was 0.7%. Extracted electric fields (along the dotted line) and the surface contour lines are shown on the subject's T_1 image in Fig. 3.

For the target point located in the DLFPC, there was an average difference of 0.8 V/m in magnitude and a maximum difference of 50 V/m between the *mri2mesh* and *headreco* segmentation methods. The average percentage difference was 2.8%.

Fig. 3 (continued) field results and the thin contours show the *headreco* field results (*fshires* contours are not shown because they are almost identical to the *mri2mesh* results). Color indications for surfaces are red for skin, orange for skull, yellow for CSF, cyan for gray matter, and purple for white matter. The dotted white line on the axial cross section is a projection onto the XY plane of the 100 mm line running along the axis perpendicular from the coil. The electric field result along the dotted white line was extracted as shown at the bottom of the figure. The dotted black line indicates the location of the target point of stimulation. The field peaks were observed at the anatomical structure transition points (represented by arrows): (1) Skin-Skull, (2) Skull-CSF, (3) CSF-GM, (4) GM-WM

Axial Cross Section

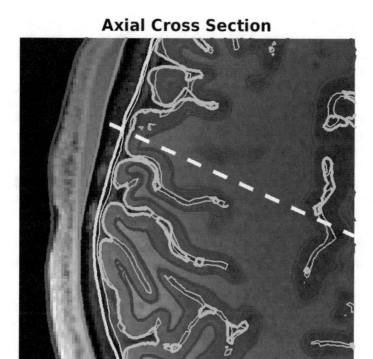

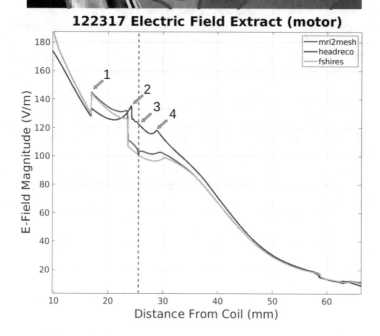

Fig. 3 Example of electric field targeted on motor cortex is shown along with an axial cross section of a subject (target stimulation point in magenta). The thick contours show the *mri2mesh* field

In the region of interest within 5 mm of the target point, the average percentage difference was approximately 1.8%. For the *mri2mesh* and *fshires* segmentation comparison, the average percentage difference in the region of interest surrounding the target point was 0.1%. Extracted electric fields (along the dotted line) and the surface contour lines are shown on the subject's T_1 image in Fig. 4.

The electric field difference across all 16 subjects at the target point between *headreco* and *mri2mesh* was statistically significant ($p = 0.005$) for the motor cortex and not significant between *mri2mesh* and *fshires* ($p = 0.83$). The electric field difference at the target point for the DLFPC was not significant for either the comparison between *mri2mesh* and *headreco* ($p = 0.19$) or between *mri2mesh* and the *fshires* ($p = 0.23$). The average percentage differences in the electric field over the 100 mm line for all subjects and target points are shown in Tables 1 and 2.

The average surface mappings for both the frontal and motor cortices are shown in Figs. 5 and 6. The average difference between the *headreco* and *mri2mesh* electric field results for the motor cortex are shown in Fig. 7. The electric fields and corresponding differences are mapped onto the *fsaverage* inflated surface along with cortical parcellation contours.

4 Discussion and Conclusion

This study focused on an evaluation of the electric field differences between The surface displacement between *mri2mesh* and *headreco* were seen at the CSF boundary, where the CSF surface was estimated closer to the gray matter for the *headreco* segmentation. Such surface displacement aligns with results shown in previous studies (cf., [11, 13]). In particular, Seiger et al. 2018 demonstrated that Freesurfer was more accurate in its calculation of cortical thickness; however, CAT12 based methods (such as *headreco*) were faster and yielded reliable results [16].

Differences between the *mri2mesh* and *fshires* were subtle as the underlying algorithm within FreeSurfer to segment the T_1 image is same for both methods. The lack of significant differences seen in the electric.

field between the two methods indicates that additional proceces to use a native submillimeter resolution is not necessary for BEM-FMM based computational modeling. Nevertheless, this may not be the case when highly detailed submillimeter surfaces are needed such as a very small structure.

In both the motor cortex and the DLPFC, the low average percent difference in the electric field suggests that the effect of the segmentation method differences was

---

Fig. 4 (continued) because they are almost identical to the *mri2mesh* results). Color indications for surfaces are red for skin, orange for skull, yellow for CSF, cyan for gray matter, and purple for white matter. The dotted white line on the axial cross section is a projection onto the XY plane of the 100 mm line running along the axis perpendicular from the coil. The electric field result along the dotted white line was extracted as shown at the bottom of the figure. The dotted black line indicates the location of the target point of stimulation. The field peaks were observed at the anatomical structure transition points (represented by arrows): (1) Skin-Skull, (2) Skull-CSF, (3) CSF-GM, (4) GM-WM

Axial Cross Section

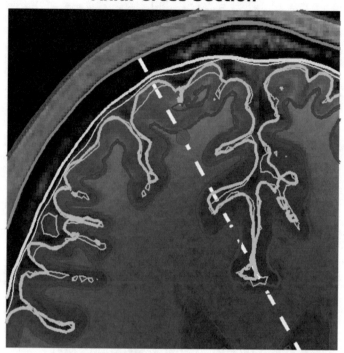

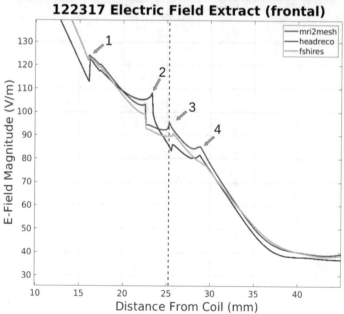

Fig. 4 Example of electric field targeted on DLPFC cortex is shown along with an axial cross section of a subject (target stimulation point in magenta). The thick contours show the *mri2mesh* field results and the thin contours show the *headreco* field results (*fshires* contours are not shown

Table 1 Average percentage difference between *mri2mesh* and *headreco* in the electric field magnitude for target points in the motor cortex and DLPFC for the 100 mm line and the average percentage difference in the region surrounding the target stimulation point (5 mm on either side of the line from the target point). Average percentage difference ranges from 82% to 112%

Subject	1	2	3	4	5	6	7	8	9	10	11	12	13	14	15	16
Motor (Obs line Avg)	1.02	0.98	1.00	1.03	0.96	0.97	0.97	0.95	0.99	1.04	1.01	0.97	0.99	0.92	0.99	0.94
Motor (target)	1.04	1.07	1.12	0.90	0.89	0.90	0.91	1.00	0.89	0.96	1.03	0.90	0.94	0.90	0.89	0.82
DLPFC (Obs line Avg)	0.97	1.08	1.07	0.95	1.03	1.00	1.01	1.08	1.12	1.04	1.04	1.06	1.01	1.03	0.99	0.96
DLPFC (target)	0.94	0.98	1.03	0.95	1.05	0.99	0.92	0.93	1.11	1.09	1.00	1.01	0.96	0.92	0.91	0.95

Table 2 Average percentage difference between *mri2mesh* and *fshires* in the electric field magnitude for target points in the motor cortex and DLPFC for the 100 mm line and the average percentage difference in the region surrounding the target stimulation point (5 mm on either side of the line from the target point). Average percentage difference ranges from 91% to 121%

Subject	1	2	3	4	5	6	7	8	9	10	11	12	13	14	15	16
Motor (Obs line Avg)	1.01	1.05	1.20	1.16	1.14	1.11	1.04	1.06	0.95	0.98	0.97	1.05	0.90	1.02	1.07	1.05
Motor (target)	0.96	0.97	1.02	1.06	1.03	1.04	1.00	1.00	1.00	0.96	0.99	1.04	0.96	0.97	1.02	0.96
DLPFC (Obs line Avg)	0.91	1.04	1.14	0.94	1.05	0.95	1.08	0.90	1.05	1.06	1.01	1.11	1.20	1.04	0.95	1.21
DLPFC (target)	0.99	1.00	1.05	0.99	1.03	0.98	0.98	1.00	1.02	1.01	0.99	1.00	0.96	0.97	0.98	0.95

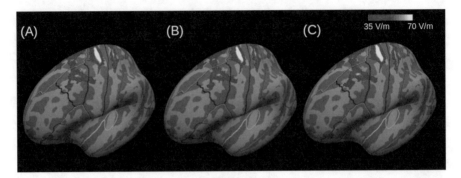

Fig. 5 Mean electric field values are mapped on the inflated surface for the motor cortex stimulation across all 16 subjects for *mri2mesh* (**a**), *fshires* (**b**), and *headreco* (**c**) along with the cortical parcellation contours. All three methods showed high electric field values in the target (precentral gyrus) along with the postcentral and caudal middle frontal gyri. No notable differences were observed between *fshires* and mri2mesh, and higher electric field values were observed in all three regions for the *headreco* method

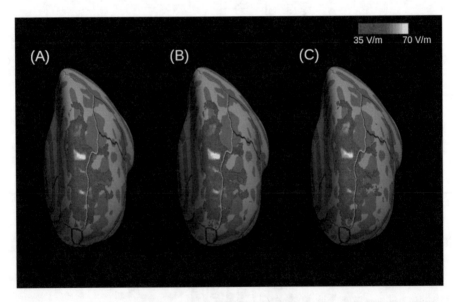

Fig. 6 Mean electric field values are mapped on the inflated surface for the DLPFC stimulation across all 16 subjects for *mri2mesh* (**a**), *fshires* (**b**), and *headreco* (**c**) along with the cortical parcellation contours. All three methods showed high electric field values in the target (superior frontal gyrus). No notable differences were observed between *fshires* and *mri2mesh*, and slightly lower electric fields were observed in all three regions for the *headreco* method

minimal. Although the overall difference was low (¡ 5%), the localized field difference near the target point across all 16 subjects was statistically significant for the motor cortex and could affect the intended stimulation there. The field difference for the DLPFC was not significant and the fields were more similar between *headreco* and *mri2mesh* as shown in Fig. 4.

Motor Cortex Difference Mapping

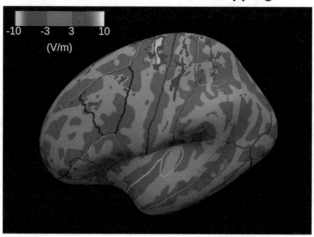

Fig. 7 Mean electric field differences between *mri2mesh* and *headreco* on the inflated surface are shown for the motor cortex stimulation. The area where the *headreco* electric field was lower than the *mri2mesh* electric field is shown in blue, and the area where the *headreco* electric field was higher than the *mri2mesh* electric field is shown in red

The percent difference in the electric field for the DLPFC was similar to that of the motor cortex along the entire 100 mm line for all 16 subjects. However, the region directly surrounding the target point showed lower variability for the DLPFC target compared to the motor cortex (Fig. 4). This trend is consistent in the average percent difference in the 5 mm region surrounding the target point. Therefore, there may be more segmentation variability in the motor cortex.

Analysis of the extracted electric field in Figs. 3 and 4 showed sudden changes in the field that resulted from the anatomical structure transitions. Some peaks are not aligned at the distance from the coil. For example, the Skull-CSF transitions in both figures were approximately 23 mm for the *headreco* segmentation whereas it was approximately 22.5 mm from the coil for the simulation with the *mri2mesh* and *fshires* methods. These sudden electric field changes resulting from segmentation differences can affect TMS therapy because the resulting neuronal excitation is a function of the electric field gradient rather than the electric field magnitude. Additional work that directly evaluates the gradient of the electric field will provide more insight into the effect of the segmentation on the TMS therapy.

The average electric field mapped on the *fsaverage* surface showed minimal differences between the *fshires* and *mri2mesh* segmentation methods as the differences were within 2 V/m. The electric field difference mapped on the *fsaverage* surface revealed that there were clusters that exceed 10 V/m of the electric field differences between *headreco* and *mri2mesh* (Fig. 7). The differences were also largely one-sided, with the fields from the *headreco* segmentation consistently higher than those from *mri2mesh*.

Coil positioning was critical to the electric field estimation. In this study, coil positioning was determined automatically based on the topology of the input meshes used. The pre-processing steps to select the proper coil position may differ significantly between segmentation methods.

The resulting field differences for each coil position were small but measurable. Though average percent differences observed along the 100 mm observation line were less than 5%, significant differences in the electric fields between segmentation methods were observed for the motor cortex simulation. Moreover, the field differences shown by subtle segmentation differences indicate an importance of patientspecific modeling as various previous studies have shown morphometric differences across age [14] and sex [17]. Future studies with different types of TMS coils and different segmentation and computational modeling methods may further improve a modeling approach for robust treatment for TMS and other neuromodulation devices.

Acknowledgments The authors would like to thank William Wartman for his assistance in visualizing the segmentations. The authors would also like to thank Drs. Brian B. Beard and Sunder S. Rajan for their helpful feedback.

Disclaimer The mention of commercial products, their sources, or their use in connection with the material reported herein is not to be construed as either an actual or implied endorsement of such products by the Department of Health and Human Services.

References

1. J. Ashburner, K. Friston, Image segmentation, in *Human Brain Function*, ed. by R. Frackowiak, K. Friston, C. Frith, R. Dolan, K. Friston, C. Price, S. Zeki, J. Ashburner, W. Penny, 2nd edn., (Academic Press, 2003), Headquartered in Cambridge, Massachusetts
2. A.M. Dale, B. Fischl, M.I. Sereno, Cortical surface-based analysis: I. Segmentation and surface reconstruction. *Neuroimage* **9**(2), 179–194 (1999)
3. R. Dubbioso, E. Raffin, A. Karabanov, A. Thielscher, H.R. Siebner, Centre-surround organization of fast sensorimotor integration in human motor hand area. NeuroImage **158**, 37–47 (2017)
4. B. Fischl, M.I. Sereno, A.M. Dale, Cortical surface-based analysis: II: inflation, flattening, and a surface-based coordinate system. NeuroImage **9**(2), 195–207 (1999)
5. M.I. Iacono, E. Neufeld, E. Akinnagbe, K. Bower, J. Wolf, I.V. Oikonomidis, D. Sharma, B. Lloyd, B.J. Wilm, M. Wyss, et al., MIDA: a multimodal imaging-based detailed anatomical model of the human head and neck. PLOS One **10**(4), e0124126 (2015)
6. K. Kazemi, N. Noorizadeh, Quantitative comparison of SPM, FSL, and brainsuite for brain MR image segmentation. J. Biomed. Phys. Eng. **4**(1), 13 (2014)
7. S. Makarov, W. Wartman, G. Noetscher, K. Fujimoto, T. Zaidi, E. Burnham, M. Daneshzand, A. Nummenmaa, Degree of improving tms focality through a geometrically stable solution of an inverse tms problem. NeuroImage **241**, 118437 (2021)
8. S.N. Makarov, G.M. Noetscher, T. Raij, A. Nummenmaa, A quasi-static boundary element approach with fast multipole acceleration for high-resolution bioelectromagnetic models. IEEE Trans. Bio-med. Eng. **65**(12), 2675–2683 (2018)
9. S.N. Makarov, W.A. Wartman, M. Daneshzand, K. Fujimoto, T. Raij, A. Nummenmaa, A software toolkit for TMS electric-field modeling with boundary element fast multipole method: an efficient MATLAB implementation. J. Neural Eng. **17**(4), 046023 (2020)

10. J.D. Nielsen, K.H. Madsen, O. Puonti, H.R. Siebner, C. Bauer, C.G. Madsen, G.B. Saturnino, A. Thielscher, Automatic skull segmentation from MR images for realistic volume conductor models of the head: assessment of the state-of-the-art. NeuroImage **174**, 587–598 (2018)
11. O. Puonti, G.B. Saturnino, K.H. Madsen, A. Thielscher, Value and limitations of intracranial recordings for validating electric field modeling for transcranial brain stimulation. NeuroImage **208**, 116431 (2020)
12. E. Raffin, G. Pellegrino, V. Di Lazzaro, A. Thielscher, H.R. Siebner, Bringing transcranial mapping into shape: sulcus-aligned mapping captures motor somatotopy in human primary motor hand area. NeuroImage **120**, 164–175 (2015)
13. R. Righart, P. Schmidt, R. Dahnke, V. Biberacher, A. Beer, D. Buck, B. Hemmer, J. Kirschke, C. Zimmer, C. Gaser, et al., Volume versus surface-based cortical thickness measurements: a comparative study with healthy controls and multiple sclerosis patients. PLOS One **12**(7), e0179590 (2017)
14. D.H. Salat, R.L. Buckner, A.Z. Snyder, D.N. Greve, R.S. Desikan, E. Busa, J.C. Morris, A.M. Dale, B. Fischl, Thinning of the cerebral cortex in aging. Cereb. Cortex **14**(7), 721–730 (2004)
15. G. Saturnino, A. Antunes, J. Stelzer, A. Thielscher. Simnibs: a versatile toolbox for simulating fields generated by transcranial brain stimulation. In *21st Annual Meeting of the Organization for Human Brain Mapping (OHBM 2015)*, 2015
16. R. Seiger, S. Ganger, G.S. Kranz, A. Hahn, R. Lanzenberger, Cortical thickness estimations of FreeSurfer and the CAT12 toolbox in patients with Alzheimer's disease and healthy controls. J. Neuroimaging **28**(5), 515–523 (2018)
17. E.R. Sowell, B.S. Peterson, E. Kan, R.P. Woods, J. Yoshii, R. Bansal, D. Xu, H. Zhu, P.M. Thompson, A.W. Toga, Sex differences in cortical thickness mapped in 176 healthy individuals between 7 and 87 years of age. Cereb. Cortex **17**(7), 1550–1560 (2007)
18. D.L. Tudorascu, H.T. Karim, J.M. Maronge, L. Alhilali, S. Fakhran, H.J. Aizenstein, J. Muschelli, C.M. Crainiceanu, Reproducibility and bias in healthy brain segmentation: comparison of two popular neuroimaging platforms. Front. Neurosci. **10**, 503 (2016)
19. D.C. Van Essen, K. Ugurbil, E. Auerbach, D. Barch, T.E. Behrens, R. Bucholz, A. Chang, L. Chen, M. Corbetta, S.W. Curtiss, et al., The human connectome project: a data acquisition perspective. NeuroImage **62**(4), 2222–2231 (2012)
20. M. Windhoff, A. Opitz, A. Thielscher. *Electric Field Calculations in Brain Stimulation Based on Finite Elements: An Optimized Processing Pipeline for the Generation and Usage of Accurate Individual Head Models*. Technical report, Wiley Online Library, 2013
21. N. Zaretskaya, B.R. Fischl, M. Reuter, J. Renvall, J. Polimeni, Advantages of cortical surface reconstruction using submillimeter 7 t memprage. NeuroImage **165**, 11–26 (2018)
22. Y. Zhang, M. Brady, S. Smith, Segmentation of brain MR images through a hidden Markov random field model and the expectation-maximization algorithm. IEEE Trans. Med. Imaging **20**(1), 45–57 (2001)

Angle-Tuned Coil: A Focality-Adjustable Transcranial Magnetic Stimulator

Qinglei Meng, Hedyeh Bagherzadeh, Elliot Hong, Yihong Yang, Hanbing Lu, and Fow-Sen Choa

1 Introduction

Transcranial magnetic stimulation (TMS) has been approved by the Food and Drug Administration (FDA) for treatment-resistant major depression [1] and Obsessive-Compulsive Disorder [2]. Its therapeutic effects in other psychiatric and neurological disorders, including drug addiction, are emerging [3, 4]. From both clinical and basic neuroscience perspectives, there has been a strong demand for stimulation tools that can reach deep brain regions with small size targeted stimulations. For example, decades of neuroimaging studies have identified malfunction of dorsal anterior cingulate cortex, insular and amygdala in a range of psychiatric disorders. These structures are 4 cm or more below the scalp. Unfortunately, with current technologies, the stimulation targets are limited to superficial brain regions, or otherwise wide brain areas are stimulated when a deep brain structure is targeted.

Hedyeh Bagherzadeh contributed equally with all other contributors.

Q. Meng (✉)
Athinoula A. Martinos Center for Biomedical Imaging, Department of Radiology, Harvard Medical School, Massachusetts General Hospital, Charlestown, MA, USA
e-mail: QMENG1@mgh.harvard.edu

H. Bagherzadeh · F.-S. Choa
Department of Computer Science and Electrical Engineering, University of Maryland, Baltimore County, MD, USA

E. Hong
Department of Psychiatry, University of Maryland, School of Medicine and Maryland Psychiatric Research Center, Baltimore, MD, USA

Y. Yang · H. Lu
Intramural Research Program, National Institute on Drug Abuse (NIDA), National Institutes of Health (NIH), Baltimore, MD, USA

© The Author(s) 2023
S. Makarov et al. (eds.), *Brain and Human Body Modelling 2021*,
https://doi.org/10.1007/978-3-031-15451-5_6

89

The output of a TMS coil can be treated as field emission from a finite size aperture and follows a specific depth-focality tradeoff rule. Deng et al. theoretically calculated the depth-focality profiles of 50 TMS coils [5]. Two groups of mainstream coils, circular and the figure-8, formed two depth-focality tradeoff curves, respectively. The study concluded that at shorter depth, which is smaller than 3.5 cm, all figure-8 type of coils follows a better depth-focality tradeoff rule and it will be advantageous to use the figure-8 coil. A number of studies have attempted to design TMS coils for enhanced the penetration depth or improved the focality. Rastogi modified conventional figure-8 coil to improve its focality but that, on the other hand, significantly weakened the electric field strength generated in brain tissues [6]. Crowther suggested a "Halo coil" design to improve the penetration depth of the conventional circular coils [7, 8], but this design sacrifices the coil's focality. Luiz modeled multi-channel coil arrays to improve the focality and penetration depth profile [9]. However, this design involved complicated coil structures, and it required higher efficiency of the coils' cooling system. Alternative coil design strategy is needed to go beyond the depth-focality tradeoff limitation. Roth et al. have developed the H-coil for human deep brain stimulation, but the design is still limited by the depth-focality tradeoff with a relatively large field spread [5, 10, 11]. We recently reported a multi-layer winding-tilted coil design approach for focused rodents' brain stimulation that induced unilateral movements [12]. The goal of this study is to extend this novel strategy to design TMS coils for deep and focused human brain stimulation and compare its depth-focality characteristic with other conventional coils.

2   Methods

Figure 1a illustrates our TMS coil design (see reference [12] for details). In this study, the coil dimensions have been extended for human brain stimulation. For each single circular coil, the inner and outer diameters are 8 cm and 9 cm, respectively. The thickness of each single coil is 1 cm and 5 circular coils are accumulated along the central axis in this model. Considering the overall inductance of the coil model, we further extended the coil length while using narrow (lower value of 'd_o–d_i') but thick (higher value of h) coil windings, so that the total turn number could be low enough to control the coil inductance.

The head model is a homogenous sphere with the diameter of 17 cm and isotropic electrical conductivity of 0.33 S/m^{-1}, The definitions of stimulation depth and focality in Fig. 1b are based on the half-value depth ($d_{1/2}$) and half-value volume ($V_{1/2}$). Modeling frequency was 5 kHz All conditions are identical to the modeling in the study by Deng et al. [5], except that the software we used for calculation is the COMSOL AC/DC module (finite element analysis software, COMSOL Inc.), which is different from that in their study. To calibrate our calculation using COMSOL, we firstly selected 4 coils, which were already documented in the study by Deng et al., and compared our results with theirs. The selected coil models for calculation calibration were the 50 mm, 70 mm and 90 mm circular and the 70 mm figure-8 Magstim coils.

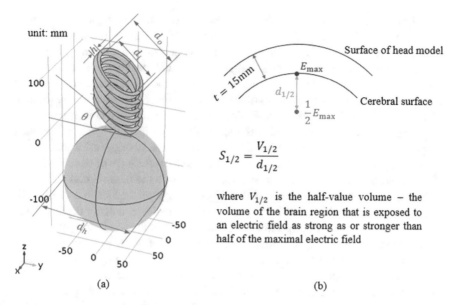

where $V_{1/2}$ is the half-value volume – the volume of the brain region that is exposed to an electric field as strong as or stronger than half of the maximal electric field

$$S_{1/2} = \frac{V_{1/2}}{d_{1/2}}$$

Fig. 1 (**a**) Simulation model and induced electric field distribution in the human head model; (**b**) The definitions of the stimulation depth and focality

Table 1 Comparison of the stimulation depth ($d_{1/2}$) and focality ($S_{1/2}$) calculated by COMSOL with data recorded in Deng's study

Coil type	50 mm circular magstim coil	70 mm circular magstim coil	90 mm circular magstim coil	70 mm figure-8 magstim coil
$d_{1/2}$ (by COMSOL) / cm	1.31	1.49	1.75	1.45
$d_{1/2}$ (by Deng et al.) / cm	1.29	1.44	1.74	1.41
$S_{1/2}$ (by COMSOL) / cm²	53.1	65.8	87.8	13.8
$S_{1/2}$ (by Deng et al.) / cm²	53.7	66.0	87.4	14.8

3 Results

3.1 Cross-Validation of Theoretical Simulation

The stimulation depth and focality calculated by COMSOL of the 4 selected coils (50 mm, 70 mm and 90 mm circular and the 70 mm figure-8 Magstim coils) as calibrations are listed in Table 1, compared with the data in Deng's study [5]. Both the stimulation depth and focality from the two finite element analysis software matched reasonably well with minor differences.

Theoretical simulation of multi-layer winding tilted coil design.

For our multi-layer winding tilted coil design, we firstly investigated how the tilt angle θ affected the stimulation depth and focality. The angle was adjusted from 0 degree to 70 degrees with a step of 10 degrees without changing any other parameters in the modeling. The values of the stimulation depth and focality were marked in the same plot summarized in Deng et al. in Fig. 2 [5]. Coils with 0-degree tilt angle (or flat coils) are located along the circular coil curve. As the tilt angle increases, the location of the coil in the depth-focality tradeoff plot moves from the circular coil curve towards the figure-8 coil curve. If the number of the winding layer is set to 5, the coil location drops on the figure-8 coil curve when the tilt angle reaches 50 degrees. Further enlarging the tilt angle enables the curve, which is plotted from the trace of our coil designs (formed by the blue square dots in Fig. 2), to penetrate the figure-8 coil curve. For example, when the tilt angle is adjusted to 60 or 70 degrees, the locations of the coils in the plot are below the figure-8 coil curve, and that indicates a better depth-focality characteristics than the existing TMS coils. The number of the winding layers are also adjusted to 2 and 9. We finalize that a smaller number of winding layers moves the curve of the coil in the plot towards the left side. For a certain tilt angle, both the stimulation depth and focality decrease. However, when the winding layer number increases from 5 to 9, the stimulation depth is not considerably improved. This phenomenon may be cause by the longer distance from the stimulation target to the few top layers, which significantly weakens the electric field strength at the stimulation target contributed by those layers. Figure 3 presents the induced electric field distribution on the human head model surface by the proposed coil designs with various design parameters.

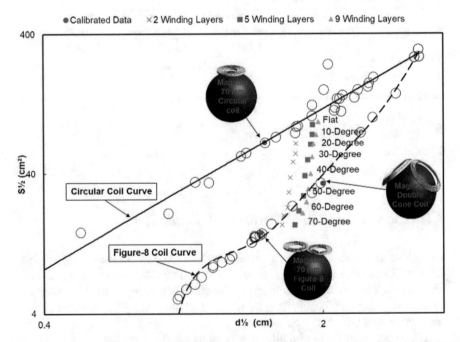

Fig. 2 Calculations of stimulation depth and focality for the multi-layer winding-tilted coil design with air core and tilt angle ranging from 0 to 70 degrees (green curve), the number of winding layers of 2, 5 and 9

To further demonstrate the advantage of our coil design, we compared the elec-

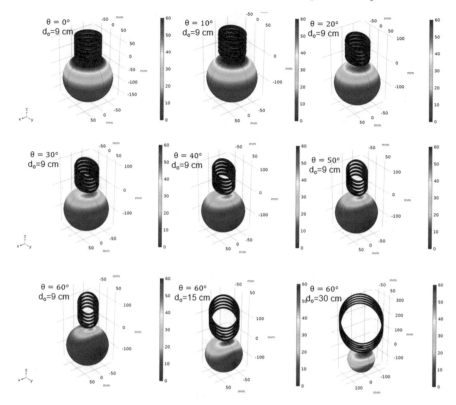

Fig. 3 Induced electric field distribution on the human head model surface by coil designs with different design parameters

tric field decay rate in the brain model for 6 coil structures, as shown in Fig. 4, and experimentally demonstrated the tilt angle could improve the focality of the induced electric field distribution. The electric field decay rate curves in the human head model in Fig. 4 indicate that neither the application of ferromagnetic core nor the tilt angle θ is able to improve the field decay rate, but the winding accumulation along the coil's central axis considerably improves it. For example, at the depth of 3 cm, a 5-layer winding accumulation improves the electric field decay rate by 4–5%.

3.2 Experimental Validation

To verify how the angle θ affected the focality, we fabricated 3 coil prototypes with winding's tilt angles of 20, 10 and 0 degree. They shared the same coil length, inner and outer diameters, which were 4.4 cm, 3.8 cm and 7.5 cm respectively. Each coil

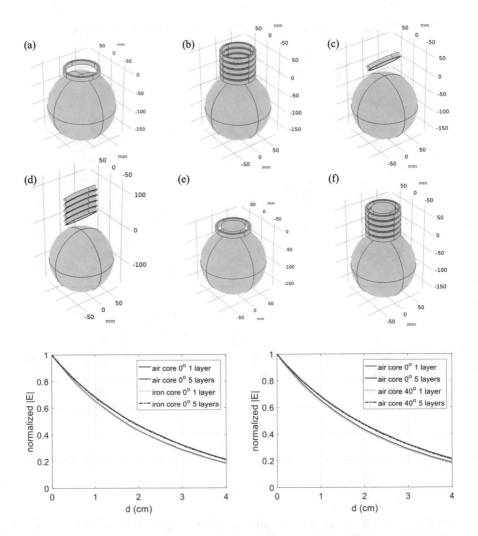

Fig. 4 Comparison of field decay rates for 6 different coil structure models: (**a**) single-layer flat circular coil with air core; (**b**) 5-layer flat circular coil with air core; (**c**) single-layer circular coil with air core and 40-degree tilt angle; (**d**) 5-layer circular coil with air core and 40-degree tilt angle; (**e**) single-layer flat circular coil with ferromagnetic core; (**f**) 5-layer flat circular coil with ferromagnetic core

was wrapped by 20 turns of the litz wires, and each turn contained a bundle of 135 piece of AWG30 wires. The TMS coil was driven by a customized driving circuit, in which an insulated gate bipolar transistor (IGBT) was used as the switch to control TMS pulses and a capacitor bank charged by a power supply was the main current source [13]. The charging voltage of the capacitor bank was set to 100 V and the pulse duration was 250 µs. The induced electric field was measured with a modified Rogowski coil electric field probe customized in our lab [14]. The electric field was mapped in the medium of air within a plane 2 cm away from the coil surface. Since

At 2cm depth:

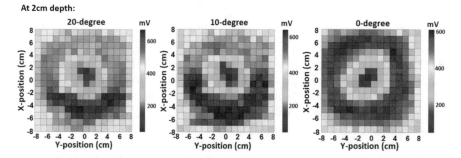

Fig. 5 Induced electric field measurements using modified Rogowski coil probe for the 3 coil prototypes with 20 degrees, 10 degrees and 0 degree wire wrapping tilt angle at the depth of 2 cm in air medium

the electric field components along the Z axis (in parallel with the coil's central axis) was small enough to be negligible, only the X and Y components of the electric field were measured. Considering the probe size, we mapped the electric field at a step size of 5 mm within the X-Y plane. An area of 8 cm × 8 cm was scanned for each coil.

Figure 5 shows the heat maps of the electric field distributions for the 3 scanned coils. For the coil with a tilt angle of 20 degrees, the area with the field strength over or equal to 80% of its peak value (E_{peak}) is only 14.5 cm^2; while for the other 2 coils, the values have reached 22.5 cm^2 (for 10-degree tilt) and 30.75 cm^2 (for 0-degree tilt), respectively. It is also found that a larger tilt angle of the coil windings would slightly increase the coil's inductance. For example, the inductances of the 3 scanned coils were measured to be 52.2 μH, 51.21 μH and 49.4 μH, respectively.

4 Discussions

The combination of air-core accumulated windings along the coil's central axis and the tilt angle of the windings provides a significant innovation to the depth-focality profile of TMS coils. The tilt angle technique is a mild symmetry breaking method. It does not reduce the equivalent field emission aperture size or speed up the electric field decay rate along the coil's central axis direction, but significantly distorts the ring shape electric field distribution, resulting in a much smaller focal spot. The only limitation of this coil design, to our knowledge, is its possible larger inductance compared with conventional TMS coils. The current flowing inside the coil I(t) can be expressed as.

$$I(t) = \frac{Vc}{\omega L} \sin(\omega t) \exp(-\sigma t), \qquad (1)$$

where ω and σ are related to the charging of the capacitors (C) in the stimulator circuit, the inductance of the coil (L), and the resistance in the LC circuit. V_c is the voltage to charge the capacitors [15]. The induced electric field $E(t)$ can be expressed as

$$E(t) = \alpha \frac{dI}{dt} \cong \alpha \frac{Vc}{L}. \tag{2}$$

So, E is inversely proportional to L. The Magstim human TMS coils always have an inductance of around 20 μH (from the Magstim Rapid [2] system manual) [16]. Our prototype human TMS coil design may have a higher inductance due to its length. This would require a stimulator of higher power output to drive the coil and enhance the current load in it, and on the other hand, the turn number of the coil can be reduced to limit its inductance to a reasonable value.

The curves we plotted for our multi-layer winding-tilted coil design in the depth-focality tradeoff profile in Fig. 2 demonstrate better focality than the H coils. The half-value depth values are comparable with the H coil designs. The half-value depth values are larger than the conventional figire-8 coils. Our coil design has achieved better depth-focality characteristic than a large sized double cone coil for human brain stimulation [5]. However, the double cone coil is known to induce scalp and facial pain and has limitations on its stimulation targeting site due to its geometry [17]. Our coil design has the advantage of much smaller contact area with the human head during the stimulation, and that may reduce or even exempt the scalp or facial pain. Moreover, with our coil design, it is feasible for users to conduct multisite stimulation, which cannot be accomplished by the double cone coils. Our design has provided the current best method for multisite human brain stimulation than all the conventional TMS coils considering both the stimulation depth and focality. The stimulation depth and focality are adjustable by the geometry of the coil design, for example, by tuning the tilt angle of the windings and the coil's outer diameter. This novel design provides a promising solution for the future deep and focused multisite human brain stimulation.

Acknowledgement The research is supported by the intramural research program of NIDA, NIH, NSF grant ECCS-1631820, NIH grants MH112180, MH108148, MH103222, and a Brain and Behavior Research Foundation grant.

Declaration of Competing Interest The authors declare that there is no conflict of interest for this manuscript.

References

1. J.P. O'Reardon, H.B. Solvason, P.G. Janicak, S. Sampson, K.E. Isenberg, Z. Nahas, W.M. McDonald, D. Avery, P.B. Fitzgerald, C. Loo, M.A. Demitrack, Efficacy and safety of transcranial magnetic stimulation in the acute treatment of major depression: a multisite randomized controlled trial. Biol. Psychiatry **62**(11), 1208–1216 (2007)
2. R. Voelker, Brain stimulation approved for obsessive-compulsive disorder. JAMA **320**(11), 1098 (2018)
3. E. Bellamoli, P. Manganotti, R.P. Schwartz, C. Rimondo, M. Gomma, G. Serpelloni, rTMS in the treatment of drug addiction: an update about human studies. Behav. Neurol. **2014**, 815215 (2014)

4. M.S. Barr, F. Farzan, V.C. Wing, T.P. George, P.B. Fitzgerald, Z.J. Daskalakis, Repetitive transcranial magnetic stimulation and drug addiction. Int. Rev. Psychiatry **23**(5), 454–466 (2011)
5. Z.D. Deng, S.H. Lisanby, A.V. Peterchev, Electric field depth–focality tradeoff in transcranial magnetic stimulation: Simulation comparison of 50 coil designs. Brain Stimul. **6**(1), 1–3 (2013)
6. P. Rastogi, E.G. Lee, R.L. Hadimani, D.C. Jiles, Transcranial magnetic stimulation-coil design with improved focality. AIP Adv. **7**(5), 056705 (2017)
7. L.J. Crowther, P. Marketos, P.I. Williams, Y. Melikhov, D.C. Jiles, J.H. Starzewski, Transcranial magnetic stimulation: improved coil design for deep brain investigation. J. Appl. Phys. **109**(7), 07B314 (2011)
8. Y. Meng, R.L. Hadimani, L.J. Crowther, Z. Xu, J. Qu, D.C. Jiles, Deep brain transcranial magnetic stimulation using variable "Halo coil" system. J. Appl. Phys. **117**(17), 17B305 (2015)
9. L. Gomez, F. Cajko, L. Hernandez-Garcia, A. Grbic, E. Michielssen, Numerical analysis and design of single-source multicoil TMS for deep and focused brain stimulation. IEEE Trans. Biomed. Eng. **60**(10), 2771–2782 (2013)
10. Y. Roth, A. Zangen, M. Hallett, A coil design for transcranial magnetic stimulation of deep brain regions. J. Clin. Neurophysiol. **19**(4), 361–370 (2002)
11. Y. Roth, A. Amir, Y. Levkovitz, A. Zangen, Three-dimensional distribution of the electric field induced in the brain by transcranial magnetic stimulation using figure-8 and deep H-coils. J. Clin. Neurophysiol. **24**(1), 31–38 (2007)
12. Q. Meng, L. Jing, J.P. Badjo, X. Du, E. Hong, Y. Yang, H. Lu, F.S. Choa, A novel transcranial magnetic stimulator for focal stimulation of rodent brain. Brain Stimulation: Basic, Translational, and Clinical Research in Neuromodulation. 2018 May 1;11(3):663–5.
13. Meng Q, Loiacono J, Choa FS, Application of insulated gate bipolar transistor in transcranial magnetic stimulation system development. In: International Semiconductor Device Research Symposium (ISDRS) 2016 Dec
14. Q. Meng, M. Daugherty, P. Patel, S. Trivedi, X. Du, E. Hong, F.S. Choa, High-sensitivity and spatial resolution transient magnetic and electric field probes for transcranial magnetic stimulator characterizations. Instrumentat. Sci. Technol. **46**, 1–17 (2017)
15. A.V. Peterchev, R. Jalinous, S.H. Lisanby, A transcranial magnetic stimulator inducing near-rectangular pulses with controllable pulse width (cTMS). IEEE Trans. Biomed. Eng. **55**(1), 257–266 (2007)
16. Magstim.com. Magstim Coils. [Online]. Available at: https://www.magstim.com/product-category/coils/
17. P.M. Kreuzer, M. Schecklmann, A. Lehner, T.C. Wetter, T.B. Poeppl, R. Rupprecht, D. de Ridder, M. Landgrebe, B. Langguth, The ACDC pilot trial: targeting the anterior cingulate by double cone coil rTMS for the treatment of depression. Brain Stimul. **8**(2), 240–246 (2015)

Part IV
Low Frequency Electromagnetic Modeling and Experiment: Spinal Cord Stimulation

Interplay Between Electrical Conductivity of Tissues and Position of Electrodes in Transcutaneous Spinal Direct Current Stimulation (tsDCS)

Sofia R. Fernandes, Mariana Pereira, Sherif M. Elbasiouny, Yasin Y. Dhaher, Mamede de Carvalho, and Pedro C. Miranda

1 Introduction

Transcutaneous spinal direct current stimulation (tsDCS) is a non-invasive stimulation technique considered as a possible therapeutic resource for spinal cord dysfunctions such as chronic pain or motor system lesion [1]. tsDCS consists in the application of low intensity direct currents of 2–4 mA to the spinal cord (SC) using electrodes placed over the vertebral column near target regions. Exploratory experimental studies in humans pursued in the last decade demonstrated the neuromodulatory potential of tsDCS to change signal transmission along nociceptive ascending pathways and spinal motor and reflex circuits in cervical and thoracic spinal segments [2–5].

Yasin Y. Dhaher, Mamede de Carvalho and Pedro C. Miranda are joint last authors.

S. R. Fernandes (✉)
Instituto de Biofísica e Engenharia Biomédica, Faculdade de Ciências, Universidade de Lisboa, Lisbon, Portugal

Instituto de Fisiologia, Instituto de Medicina Molecular João Lobo Antunes, Faculdade de Medicina, Universidade de Lisboa, Lisbon, Portugal
e-mail: srcfernandes@fc.ul.pt

M. Pereira · M. de Carvalho
Instituto de Fisiologia, Instituto de Medicina Molecular João Lobo Antunes, Faculdade de Medicina, Universidade de Lisboa, Lisbon, Portugal

S. M. Elbasiouny
Department of Biomedical, Industrial and Human Factors Engineering, College of Engineering and Computer Science, Wright State University, Dayton, OH, USA

Department of Neuroscience, Cell Biology and Physiology, Boonshoft School of Medicine and College of Science and Mathematics, Wright State University, Dayton, OH, USA

© The Author(s) 2023
S. Makarov et al. (eds.), *Brain and Human Body Modelling 2021*,
https://doi.org/10.1007/978-3-031-15451-5_7

The effects of tsDCS rely mainly on the electric field (EF) induced in the nervous tissue, similarly to brain stimulation techniques. The EFs may contribute to inhibit or facilitate neuronal responses, by transiently changing the resting membrane potential. This effect will depend on how each spinal neuron or circuit is orientated relative to the EF. Just as in brain stimulation, the spatial distribution of the EF induced by tsDCS depends on electrode number, design (shape and structure), placement relative to target, current intensity and polarity (anodal/cathodal). Computational studies using numerical methods to predict the EF and current distribution in realistic human models may be powerful tools to fine-tune experimental tsDCS protocols targeting specific spinal regions of interest [6–9].

Modelling studies in tsDCS generally consider square or rectangular electrodes in bipolar montages, which originate EFs in the spinal cord that increase with distance between electrodes at the cost of a larger (less focal) target region. Cortical stimulation studies predict that montages using smaller electrodes in a multi-electrode setup can induce maximum EF in a smaller target region [10–12]. Thus, one question of interest is to determine if a multi-small-electrodes paradigm in tsDCS can also originate a more focal EF in the spinal cord, as in brain stimulation, which could be relevant for neuromodulation of specific spinal circuits.

The effects of different montage settings are also dependent of the electrical properties of tissues located in the current path to the stimulation target. Sensitivity analyses performed on transcranial direct current stimulation (tDCS) demonstrated changes of the EF magnitude when varying the conductivities of tissues. Scalp, CSF, gray matter (GM) and skull were the tissues that introduced larger variability on EF values in the cortex [13, 14]. The EF induced in tsDCS also show spatial characteristics, such as local hotspots, that seem to be related with anatomical features such as vertebrae edges and disks protrusions into the spinal canal. These local features may be related with an interplay between the conductivities of different tissues, such as vertebrae, intervertebral disks and CSF [7–9].

This chapter is dedicated to study the interplay between the conductivities of tissues and two different types of electrode montages on the spatial distribution of the EF induced by tsDCS. These types of montages will be modelled considering tsDCS over the lower thoraco-lumbar spinal cord: a montage with two large and square

Y. Y. Dhaher
Department of Physical Medicine and Rehabilitation, Northwestern University, Chicago, IL, USA

Department of Mechanical Engineering, University of Louisiana at Lafayette, Lafayette, LA, USA

Department of Physical Medicine and Rehabilitation, University of Texas Southwest, Dallas, TX, USA

Department of Orthopedic Surgery, University of Texas Southwest, Dallas, TX, USA

P. C. Miranda
Instituto de Biofísica e Engenharia Biomédica, Faculdade de Ciências, Universidade de Lisboa, Lisbon, Portugal

electrodes (50×50 mm^2) and a grid with 4×4 small circular electrodes ($\varnothing$ = 10 mm). For these two types of montages, EF changes will be investigated considering rotation of the grid or increases in distance between the two large electrodes. The impact of electrical conductivities of the tissues located between the electrodes and the SC will be addressed in each montage considering three different types of models: homogeneous, semi-homogeneous and heterogeneous. The clinical relevance of differences in the EF induced by the multi-electrode grid system and two-electrode montages in tsDCS application will be discussed. Furthermore, the influence of different conductivity properties of tissues will be studied since it can be determinant to optimize stimulation delivery for intended spinal targets.

2 Methods

2.1 Realistic Human Model and Electrode Placement

This study used the same realistic human model as in previous works (e.g [9]). This model was based on relevant tissue masks from the 34 years-old male Duke model of the Virtual Population Family [15], comprising 15 tissues (Fig. 1). The spinal GM was artificially designed considering general anatomical knowledge and measurements from the Visible Human Data Set (National Library of Medicine, NLM, Visible Human Project®, www.nlm.nih.gov/research/visible/visible_human.html). The full model was truncated at the level of the thighs and above the elbows, to reduce model size and computational time.

Electrodes were represented as hydrogel layers with 1 mm thickness, considering two geometries available for standard electrodes used in transcutaneous electrical nerve stimulation (TENS): 50 × 50 mm^2 square electrodes (Fig. 2a); circular

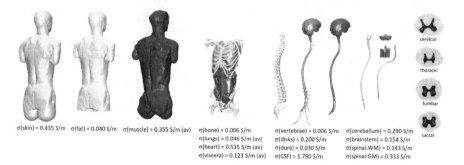

σ(skin) = 0.435 S/m   σ(fat) = 0.040 S/m   σ(muscle) = 0.355 S/m (av)   σ(bone) = 0.006 S/m   σ(vertebrae) = 0.006 S/m   σ(cerebellum) = 0.290 S/m
σ(lungs) = 0.046 S/m (av)   σ(disks) = 0.200 S/m   σ(brainstem) = 0.154 S/m
σ(heart) = 0.535 S/m (av)   σ(dura) = 0.030 S/m   σ(spinal-WM) = 0.143 S/m
σ(viscera) = 0.123 S/m (av)   σ(CSF) = 1.790 S/m   σ(spinal-GM) = 0.333 S/m

cervical
thoracic
lumbar
sacral

Fig. 1 Tissues included in the human model and corresponding isotropic values of electrical conductivity σ, as assessed in Fernandes et al. [9]. (av) indicates that conductivities were averaged over different tissues' components or longitudinal/transversal values (CSF – cerebrospinal fluid; WM – white matter; GM – gray matter)

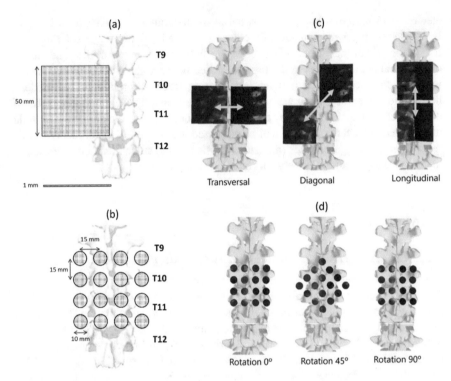

Fig. 2 Geometry and dimensions of the square (**a**) and circular (**b**) electrodes represented in the model. Variations considered for 2-electrode montages (**c**) and Grid montages (**d**); anodes and cathodes are represented by red and blue electrodes, respectively

electrodes with a diameter of 10 mm (Fig. 2b). TENS electrode design was considered in this study to adequately reproduce the electrodes that will compose the grid montages to be studied by one of our clinical teams. Two types of electrode settings were considered: bipolar montages with two square electrodes (2-electrode); multi-electrode montages with 16 circular electrodes placed in a 4×4 electrode array (Grid). Both montages were centered midway between T9 and T11 vertebrae spinous processes. The 2-electrode montage was varied in alignment relative to the vertebral column – transversal (2-T), diagonal (2-D), longitudinal (2-L) – and distance – 5 mm and 65 mm between electrode edges (Fig. 2c). The electrode Grid was rotated by 45° and 90° relative to the first position (rotation 0°, Fig. 2d). Surface meshes were optimized and assembled, and volume meshing was performed with the 3-MATIC module from MIMICS (MIMICS software, v16), resulting in $2×10^7$ tetrahedral elements for the entire human model with electrodes. Average meshing time was 4 hours per electrode montage.

Table 1 Isotropic electrical conductivity models

Model	Name	σ values
Homogeneous	Homogeneous	σ = 0.289 S/m for all tissues
Semi-homogeneous	GM	σ(GM) = 0.333 S/m; σ(other tissues) = 0.289 S/m
	WM	σ(WM) = 0.143 S/m; σ(other tissues) = 0.289 S/m
	CSF	σ(CSF) = 1.79 S/m; σ(other tissues) = 0.289 S/m
	Dura	σ(dura) = 0.030 S/m; σ(other tissues) = 0.289 S/m
	Vertebrae	σ(vertebrae) = 0.006 S/m; σ(other tissues) = 0.289 S/m
	Disks	σ(disks) = 0.200 S/m; σ(other tissues) = 0.289 S/m
Heterogeneous	Isotropic	σ values from Fig. 1

GM gray matter, *WM* white matter, *CSF* cerebrospinal fluid

2.2 Electrical Properties of Tissues and Electrodes

Tissues and hydrogel were assumed to be purely resistive with isotropic electrical conductivities. An electrical conductivity of 0.02 S/m was assigned to hydrogel in all simulations [16]. Three different levels of tissue heterogeneity were considered (Table 1):

Heterogeneous model – Tissues and hydrogel were assumed to be purely resistive with isotropic electrical conductivities, considering DC electrical tissue properties compiled in our previous work [9] (Fig. 1);

Homogeneous model – All tissues were assigned an isotropic electrical conductivity of 0.289 S/m, that corresponds to the volume-weighted average of the isotropic conductivities of each tissue included in the model;

Semi-homogeneous model – Considers the same conductivity as the homogeneous model for all tissues, except for one tissue in the vertebral column (spinal gray matter (GM), spinal white matter (WM), CSF, dura, vertebrae or disks) resulting in six different models.

2.3 Electric Field Calculations

Electric field (EF) spatial distribution was simulated with COMSOL Multiphysics using the finite element method (FEM). Total current intensity was considered as 4 mA (the maximum delivered by a typical tsDCS device): 4 mA/electrode in 2-electrode montages and 0.5 mA/electrode in Grid montages. Boundary conditions were applied according to [17], considering the hydrogel top surface as isopotential. A total of 72 simulations (9 montages variations × 8 conductivity models) were performed, with 2.2×10^7 degrees of freedom and solution time of about 30 minutes per simulation on a computer with 2 quad-core Intel® Xeon® processors clocked at 3.2 GHz and 48 GB of RAM.

Modelling studies in tDCS reported volume-average E-field values larger than 0.15 V/m over the hand knob, when reproducing clinical settings with observed neuromodulatory effects [17, 18]. tsDCS neuromodulation will be assumed if the average EF exceeds this value in the SC.

EF components were defined as 3 orthogonal vectors: $\mathbf{E_{long}}$ – caudal-rostral oriented and tangent to SC axis; $\mathbf{E_{vd}}$ – ventral-dorsal oriented and perpendicular to SC axis; $\mathbf{E_{rl}}$ – right-left oriented and perpendicular to SC axis. Spatial profiles of the magnitude of the total EF and of its components along the spinal WM and GM were determined by considering EF averages calculated at 1-mm thick axial slices along the SC length.

3   Results

3.1   Current Delivery for 2-Electrode and Grid Placements: Safety Considerations

The electrode montages considered in this study may present some safety concerns in terms of tissue damage that should be addressed: (1) in 2-electrode placements, an inter-electrode distance of 5 mm may be too small to prevent current local maxima at electrode edges; (2) small circular electrodes in Grid montage may deliver large current densities below each electrode.

Current density was predicted at skin and target tissues (spinal GM and WM) for the heterogeneous model considering all placements addressed and is summarized in Fig. 3. Current density magnitude shows hotspots in skin regions near electrode edges (Fig. 3a). For inter-electrode distances of 5 mm, these hotspots are higher between electrodes. In Grid montages, the current density is larger at electrodes' edges and maximum in skin regions located below the grid where electrodes' polarity changes from anodal to cathodal. Maximum values of the current density vary from 6.89 to 24.60 A/m^2 in skin regions near and below the electrodes, with the highest values predicted in Grid montages and the lowest values in 2-electrode montages with larger inter-electrode distances (Fig. 3b). The opposite occurs at the target tissue, where current density maxima are larger for 2-electrode montages with more distance between anode and cathode.

Current density threshold for cutaneous and nerve lesion was established to be 143 A/m^2 for DC stimulation [19]. This value has not been reached in previous tsDCS and tDCS experimental studies with no report of tissue damage [1, 5, 20, 21]. These studies applied currents of 2–4 mA using larger electrodes at larger distances. Models of these tsDCS protocols predicted current densities maxima of 12.6–21.7 A/m^2 and 0.11–0.15 A/m^2 at skin and spinal-GM, respectively [9]. These values are of

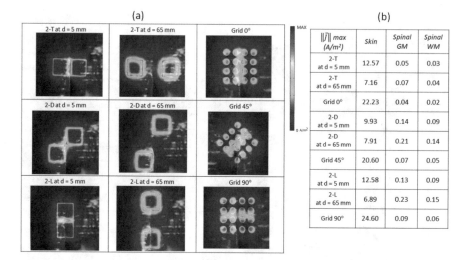

Fig. 3 (**a**) Current density magnitude distribution in skin below and near electrodes for each montage. A color scale is presented at the top right: all distributions are normalized to the respective maximum value. (**b**) Maximum values predicted for the current density magnitude in skin and spinal WM and GM for each montage

the same order of magnitude of the values predicted for the 2-electrode and Grid montages and variations presented in this study, thus the montages proposed here can be considered safe for future experimental studies.

3.2 EF Distribution for Grid and 2-Electrode Montages in the Heterogeneous Model

The heterogeneous model includes isotropic electrical conductivities for each tissue in the model (Fig. 1), thus providing a more realistic insight on the spatial characteristics of the EF at target (spinal cord) and surrounding tissues. Figure 4 shows the EF magnitude in a target volume of the spinal cord, comprising lower thoracic to lumbosacral spinal regions, from T7 to Co segments for all montages and corresponding variations in rotation, alignment and distance between electrodes. 2-electrode and Grid montages can be grouped regarding similar EF spatial profiles: transversal/rotation 0° (T0); diagonal/rotation 45° (D45); longitudinal/rotation 90° (L90). The EF magnitude distribution along the SC is wider and reaches higher EF maximum in 2-electrode montages, especially for a larger electrode distance, whereas the distribution due to 2-electrode at 5 mm is more similar to that of Grid montages (Fig. 4). The EF direction varies with montage groups, which can be seen by comparing the maximum values of the EF components, normalized to the maximum EF magnitude (E_{max}; Fig. 4, table at top right). In T0 montages, the EF is mostly transversal, with a larger right-left component of 0.96–0.98 E_{max}. In D45 and

(a)   (b)

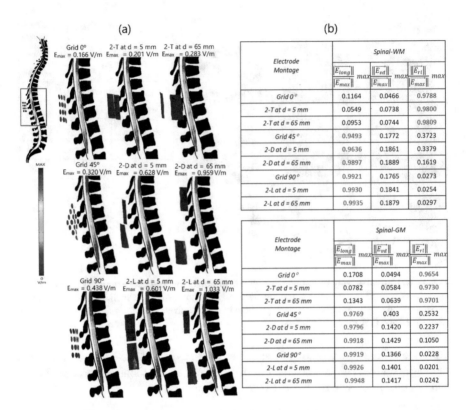

Electrode Montage	Spinal-WM		
	$\dfrac{\lVert E_{long}\rVert}{\lVert E_{max}\rVert}\,max$	$\dfrac{\lVert E_{vd}\rVert}{\lVert E_{max}\rVert}\,max$	$\dfrac{\lVert E_{rl}\rVert}{\lVert E_{max}\rVert}\,max$
Grid 0°	0.1164	0.0466	0.9788
2-T at d = 5 mm	0.0549	0.0738	0.9800
2-T at d = 65 mm	0.0953	0.0744	0.9809
Grid 45°	0.9493	0.1772	0.3723
2-D at d = 5 mm	0.9636	0.1861	0.3379
2-D at d = 65 mm	0.9897	0.1889	0.1619
Grid 90°	0.9921	0.1765	0.0273
2-L at d = 5 mm	0.9930	0.1841	0.0254
2-L at d = 65 mm	0.9935	0.1879	0.0297

Electrode Montage	Spinal-GM		
	$\dfrac{\lVert E_{long}\rVert}{\lVert E_{max}\rVert}\,max$	$\dfrac{\lVert E_{vd}\rVert}{\lVert E_{max}\rVert}\,max$	$\dfrac{\lVert E_{rl}\rVert}{\lVert E_{max}\rVert}\,max$
Grid 0°	0.1708	0.0494	0.9654
2-T at d = 5 mm	0.0782	0.0584	0.9730
2-T at d = 65 mm	0.1343	0.0639	0.9701
Grid 45°	0.9769	0.403	0.2532
2-D at d = 5 mm	0.9796	0.1420	0.2237
2-D at d = 65 mm	0.9918	0.1429	0.1050
Grid 90°	0.9919	0.1366	0.0228
2-L at d = 5 mm	0.9926	0.1401	0.0201
2-L at d = 65 mm	0.9948	0.1417	0.0242

Fig. 4 (**a**) EF distribution in a selected volume of the SC, between T7 and Co spinal segments for all montages and variations; maximum value and name of montage are on top of each plot; location of the selected volume is marked by a red rectangle over a representation of the vertebral column and colour scale for the EF magnitude are at top and middle left, respectively. (**b**) Maximum values of the EF components normalized to E_{max} in the spinal WM (top) and GM (bottom). Values of the largest component are in blue

L90, the EF direction is longitudinal, (E_{long} (max) = 0.94–0.99 E_{max}). A maximum contribution of E_{long} is present in L90 and 2-D at 65 mm. Maximum values occur mainly at T12 and L1 spinal segments, which are located midway between the electrodes or grid upper and lower borders. Figure 5 shows contributions of different montage variations inside the SC, considering an axial slice of the spinal cord at T12 level. EF similarities also translate in L0, D45 and L90 grouping at local level. D45 and L90 present different spatial patterns in the WM and GM when compared to T0. EF orientation projected in the axial plane is also represented in Fig. 5 (black arrows) and changes from group to group. D45 presents a EF axial projection with a dorsal-ventral, right-left direction, whereas L90 projection is almost only dorsal-ventral. This can also be seen when comparing values of E_{vd} and E_{rl} normalized to maximum in Fig. 4b. This orientation is consistent with anode-cathode relative position in each case.

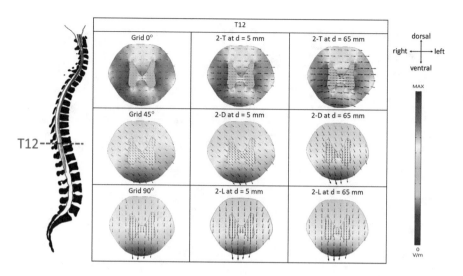

Fig. 5 EF distribution in an axial slice at T12 spinal level in spinal WM and GM for all montages. Left: location of the slice is indicated by a red dashed line over a representation of the vertebral column. Right: colour scale, with the red colour corresponding to the maximum EF in the WM in each case and orientation of the slices at the top. The black arrows have size proportional to the EF magnitude and represent the orientation of the field projected in the axial plane

Table 2 Maximum values of the EF in the spinal WM and GM in the homogeneous model, in V/m

Electrode montage	2-T		2-D		2-L		Grid rotation		
	5 mm	65 mm	5 mm	65 mm	5 mm	65 mm	0°	45°	90°
Spinal-WM	0.62	0.69	0.72	0.82	0.63	0.91	0.51	0.49	0.48
Spinal-GM	0.56	0.64	0.67	0.77	0.58	0.85	0.46	0.43	0.43

3.3 Homogeneous Model: How the Relative Positions of Electrodes Influence the EF Direction

The homogeneous model provides insight on the effect of electrode positions on the EF direction and magnitude, reducing variability that arises due to the different properties of tissues that surround the SC. 2-electrode montages originate larger EF magnitudes in the spinal WM and GM when compared to the 4×4 electrode grid. Furthermore, larger distance between the electrode edges results in an EF increase, whereas grid rotation does not change the EF considerably in terms of maximum values (Table 2). Again, 2-electrode and grid montages can be grouped regarding similar spatial profiles of the magnitude of the total EF and of its components (Fig. 6). Grid montages are more similar to 2-electrode montages with inter-electrode distance of 5 mm, compared to the montages with 65 mm. The latter also originate EF magnitude larger than 0.50 E_{max} in a wider region, by approximately two spinal segments than the grid or the 2-electrode at 5 mm.

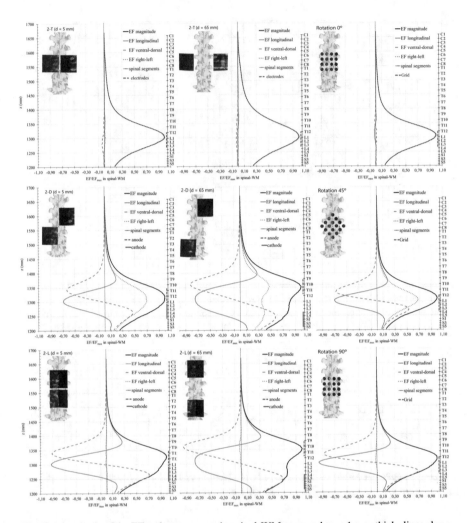

Fig. 6 Magnitude of the EF and components in spinal-WM averaged over 1-mm thick slices along the spinal cord length and normalized to EF maximum (E_{max}). The corresponding electrode montage and orientation and the legend for each element are indicated in each plot at the top left and right, respectively. The E_{rl} component is superimposed on the total EF magnitude in the first top row

The direction of the EF is consistent with the relative position of anodes and cathodes in each case. T0 montages originate an EF with a E_{rl} component that almost matches the total magnitude (Fig. 6, top row). The diagonal and longitudinal alignments (D45 and L90) increase E_{long} and E_{vd} contributions to the total EF (Fig. 6, middle and bottom rows). E_{rl} contribution is similar to the other components in D45, decreasing more in 2-D at 65 mm; this contribution disappears almost entirely at L90 montages.

The E_{long} component peaks in the anode-cathode transition region; E_{vd} presents two peaks of opposite sign, one below each anode and cathode regions. The larger distance between electrodes in 2-D and 2-L at 65 mm increases E_{vd} below the electrodes, resulting in a total EF spatial profile with two peaks, below anode and cathode regions. The anode-cathode relative placement and distance have a strong influence on the direction of the EF, allowing to establish which component can be more significant to the total EF magnitude. This could be determinant to target specific spinal neurons according to their orientation inside the SC.

3.4 Semi-homogeneous Models: How Tissues' Different Conductivities Influence the EF

The EF spatial profiles in the semi-homogeneous models can reveal the influence of the electrical conductivity of each tissue on EF components and total magnitude along the SC. The EF spatial profile is very similar to the profile for the E_{rl} component in T0 montages in the spinal WM (Fig. 7). The same type of profiles occurs in the spinal GM (not shown). This profile presents peak-like features in the spinal segments below and near the Grid or the 2-electrode. Also, the total magnitude and E_{rl} profiles mimic the semi-homogeneous vertebrae model profile, thus the conductivity of vertebrae seems to be the main factor determining the existence of the peak-like features of the EF spatial variation in the spinal WM (Fig. 7, light grey dashed line in each plot). The E_{rl} component contributes to 97–99% of E_{max} in all T0 models, which highlights the importance of the anode-cathode placement in the total EF spatial distribution (Fig. 10).

D45 montages show similar profiles for the total EF and components (Fig. 8). It presents the same type of global profiles as in the homogeneous model, where E_{long} represents 69–75% of E_{max} in the homogeneous models. This percentage is around 95–99% for the heterogeneous (isotropic) model, the larger value corresponding to 2-D at d = 65 mm (Fig. 10). This increase in E_{long} contribution should arise due to the influence of the CSF conductivity, since the total EF has a profile more similar with semi homogeneous CSF model (Fig. 8, dot line in dark grey). There are also some peak-like features in the spatial distribution at the same spinal level as in the semi-homogeneous vertebrae model, which supports the influence of the vertebrae conductivity in local hotspots. The E_{vd} component is larger in the semi-homogeneous models, presenting the same peaks below the anode and cathode regions as in the homogeneous model, but almost disappears in the heterogeneous isotropic model. This indicates that the combined effect of the different conductivities cancels out the contribution of the electrode placement to the E_{vd} component.

L90 montages have similar distributions to D45 montages, with the E_{long} component contributing to 99% of E_{max} value, and with an almost negligible E_{rl} component (Figs. 9 and 10). The E_{vd} is present in homogeneous and semi-homogeneous models, but the overall effect of different conductivities results in a less meaningful contribution in the heterogeneous model, just as seen in the D45 montages.

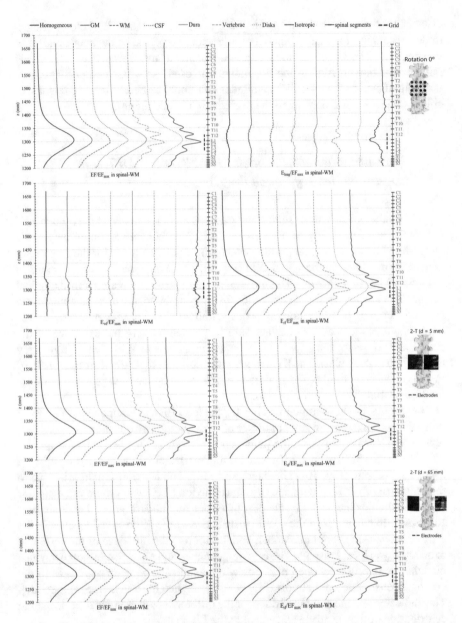

Fig. 7 EF profiles of the total and components magnitude normalized to E_{max} in the spinal-WM for T0 montages. The first two rows present the spatial profiles for the total EF, E_{long}, E_{vd} and E_{rl} for Grid 0° montage. The middle and bottom rows present the spatial profiles for the total EF and E_{rl} for Transversal 2-pads at d = 5 mm and d = 65 mm, respectively

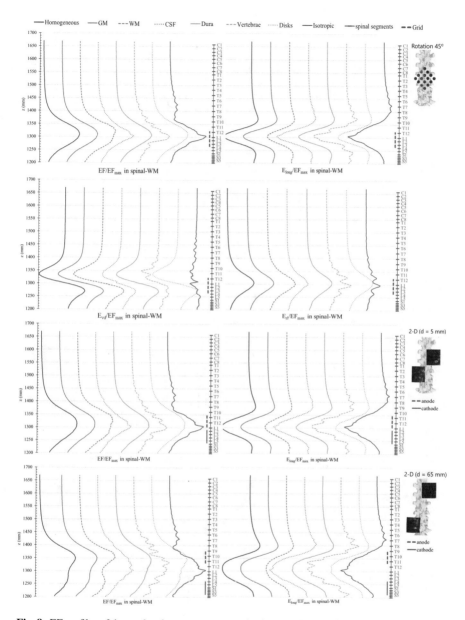

Fig. 8 EF profiles of the total and components magnitude normalized to E_{max} in the spinal-WM for D45 montages. The first two rows present the spatial profiles for the total EF, E_{long}, E_{vd} and E_{rl} for Grid 45° montage. The middle and bottom rows present the spatial profiles for the total EF and E_{long} for Diagonal 2-pads at d = 5 mm and d = 65 mm, respectively

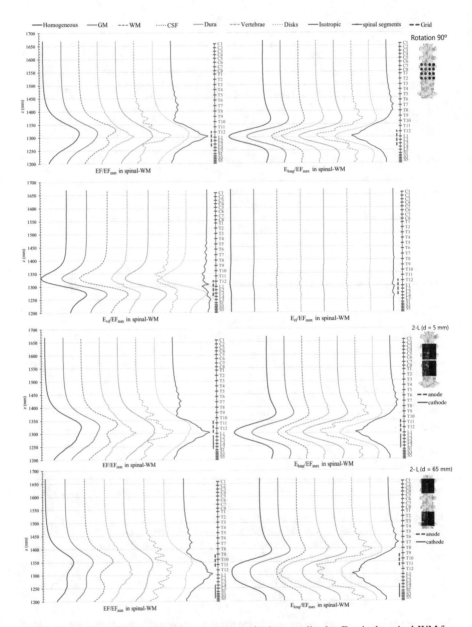

Fig. 9 EF profiles of the total and components magnitude normalized to E_{max} in the spinal-WM for L90 montages. The first two rows present the spatial profiles for the total EF, E_{long}, E_{vd} and E_{rl} for Grid 90° montage. The middle and bottom rows present the spatial profiles for the total EF and E_{long} for longitudinal 2-pads at d = 5 mm and d = 65 mm, respectively

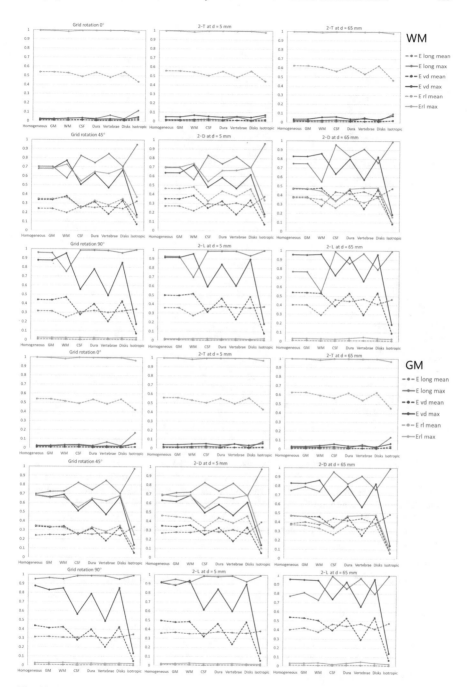

Fig. 10 Mean and maximum values of E_{long}, E_{vd}, E_{rl} normalized to E_{max} in the spinal WM and GM for all models

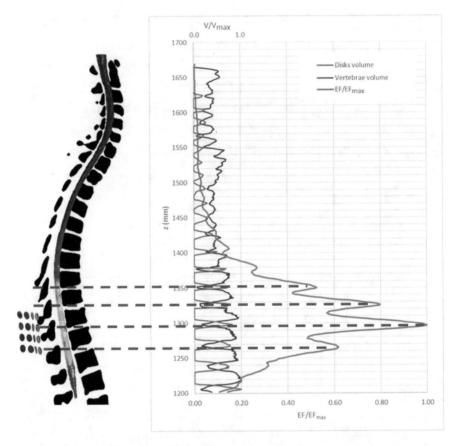

Fig. 11 Distribution of vertebral volume, disk volume and EF magnitude in Grid 0° montage, averaged over 1-mm thick slices along the SC length and normalized to maximum values in each case (horizontal bottom and upper axes correspond to EF and vertebral/disk volume, respectively). To the left, the EF magnitude volume distribution is represented with the same length for comparison and uses the same colour scale as in Fig. 4. The red lines show the spatial correspondence between local hotspots, EF peaks, disk volume maxima and vertebral volume minima

What determines the "peak-like" profile seen in the E_{rl} component? Figure 11 compares the normalized EF magnitude profile with the distribution of volume of vertebrae and disks for Grid 0° montage, where the E_{rl} component largely influences the total EF magnitude. The positions of vertebral spaces and intervertebral disks (volume minima and volume maxima, respectively) are coincident with the "peak-like" features in the EF spatial distribution. Considering that the profile of semi-homogeneous vertebrae is the only one that reflects these "peak-like" features, vertebral conductivity must have the largest influence in the formation of local hotspots.

4   Discussion

This study addresses the interplay of different electrode placements and electrical conductivity modelling paradigms in setting the main characteristics of the EF induced by tsDCS over the spinal cord. Our main finding is that the relation between the two factors is complex: anode-cathode placement determines the EF orientation, however electrical properties of tissues can change this orientation and originate local hotspots. Conversely, local hotspots that could be originated by tissue conductivity heterogeneities can disappear when the electrodes are oriented along the SC, by increasing the contribution of the highly conductive CSF. This variability in the interplay between electrodes and electrical properties of tissues in the current path is determinant to optimize EF delivery using tsDCS for a specific spinal target.

4.1   Anode-Cathode Placement Interplays with Tissues' Conductivities to Define the EF Direction

The homogeneous model is a useful tool to isolate the effect of electrode placement since it considers the human model as a completely uniform volume conductor. The main observations using this model are:

- Different electrode geometries and number are not determinant in the EF direction if the anode-cathode spatial relation is preserved – a grid of 16 small circular electrodes or a two-electrode configuration resulted in similar EF spatial profiles, if presenting the same anode-cathode relative position (Fig. 6);
- Larger distances between electrodes originate higher EF magnitudes in wider regions (Fig. 6, Table 2).

When turning to the isotropic heterogeneous model, the EF shows peak-like features and does not always preserve the spatial profiles of the EF and of its components (Figs. 4, 7, 8, and 9). For instance, E_{vd} profile in D45 and L90 homogeneous models has two main peaks and with opposite orientations below anodes and cathodes, that almost disappear in the heterogeneous model. This indicates that electrical conductivities considered in the heterogeneous (isotropic) model interfere with the EF orientation previously established by anode-cathode relative position. This raises the question: which tissues have the most impact on the EF spatial distribution in the spinal cord? Realistic numerical models of DC cortical stimulation refer CSF and skull as the tissues that can induce large variability in EF magnitude [13, 14]. Modelling studies on tsDCS revealed a negative relation between CSF volume and EF magnitude over the thoracic region when two electrodes are placed over T10 spinous process and right arm [7]. In our previous modelling study on thoraco-lumbar tsDCS, locations of vertebrae bony edges and CSF narrowing were associated with local hotspots, regardless of electrode placement [9]. The semi-homogeneous models were considered above to isolate the contribution of each tissue's conductivity. In all cases, vertebrae and CSF are the two tissues that have the most impact in local features and contribute to the global spatial profile. The EF profiles in vertebrae and CSF semi-homogeneous models closely resemble the EF in the heterogeneous model for T0 and D45/L90 montages, respectively. Vertebrae and CSF conductivity values differ by almost one order of magnitude relative to those of other tissues, which may explain a larger effect in current spreading to the surrounding tissues, thus originating the profiles observed: vertebrae has a low conductivity, which favors the passage of current through inter-vertebral space to the spinal cord, favoring the E_{rl} component; CSF's higher conductivity contributes to an increase in the longitudinal component of the EF and also contributes to current focusing effects in narrower regions originating, for instance, the sharp peak observed at T12 level in D45/L90 montages. The effect of CSF thinning was also observed in transcranial magnetic stimulation models, where the induced current flows tangential to the skull, leading to a local increase of the EF magnitude at the gyral crowns due to local CSF thinning [22].

The effect of distance was already addressed in previous tsDCS modelling works. Anode-cathode relative positions and tissues conductivities should be carefully considered when choosing a tsDCS montage. The interplay between these two factors will determine the EF orientation. The modulation of spinal neurons, just as in cortical neurons, will occur if these are aligned with the induced EF. Cellular models of spinal motor neurons identified axon terminals as the dominant cellular target of tsDCS and misplacements of 5 cm in electrodes' positions lead to a change in EF direction and a consequent reversal of polarization at target, thus the influence of electrode distance in position should be carefully considered [8]. However, DCS does not change excitability of peripheral axons, which suggests that synapses are the main targets for excitability modulation induced by DCS [23, 24]. Spinal dysfunction can have many causes but are most frequently associated with aberrant triggering of spinal reflexes and muscle tone, due to vertebral lesions, tumors or neuron degeneration. Many different neurodegenerative, inflammatory or traumatic disorders can affect spinal cord networks, with variable modulation by the same tsDCS montage. This leads to some considerations when translating modelling findings to the clinical context:

1. The selection of the topography and polarity of electrodes in tsDCS is strongly dependent on the position and orientation of the targeted spinal neurons position and their nearby anatomical frame – it is important to consider the location of the surrounding vertebrae and nearby CSF narrowing due to protruded or herniated disks;
2. Electrode grid should position the target below the anode-cathode transition region; two-electrode montages should comprise the target between the positions of the two electrodes;
3. Longitudinal/diagonal placements will be more favorable to modulate longitudinally-oriented targets; transversal placements will preferentially modulate neural targets with a mediolateral orientation;
4. In longitudinal/diagonal placements, increasing the distance between the electrodes increases the EF magnitude at the cost of focality.

4.2   Multi-electrode Montages Can Be Relevant for Differential Targeting in tsDCS

Safety concerns are a major issue when considering transcutaneous application of direct currents. Most applications of tsDCS have been more conservative, with delivery of current intensities below 3 mA. However, intensities up to 4 mA are being considered to increase neuromodulation outcomes in brain stimulation, with promising results regarding patients' tolerability [21]. The use of a multi-electrode grid may raise some issues because current density will be larger than 2-electrode montages at the skin located below the anode-cathode transition region. However, our calculations predict maximum values around one order of magnitude below the threshold for tissue damage [19]. Also, multi-electrode grid montages can reproduce the same type of EF spatial distribution than the traditional two-electrode montages, as demonstrated in this study, however with lower EF values. The maximum EF obtained in the grid rotations considered above are larger than 0.15 V/m in the heterogeneous (isotropic) model (E_{max} (grid 0°) = 0.166 V/m; E_{max} (grid 45°) = 0.320 V/m; E_{max} (grid 90°) = 0.438 V/m; Fig. 4), an assumed threshold value for effective neuromodulation (Sect. 2.3). What would be the advantage of multi-electrode grid montages, if the EF orientations reproduced with the rotations can be obtained with 2-electrode montages and with higher EF magnitude? In the clinical setting, a multi-electrode grid system may be an advantage if spinal targets with different orientations are considered, because the grid allows to obtain different anode-cathode alignments without changing the placement of the grid, for example, to have a stimulation paradigm alternating mediolateral and longitudinally oriented EFs. A grid setting using only 8 of the 16 electrodes can reproduce an EF with a spatial distribution similar to the D45 group, by considering the 4 top right electrodes as anodes and the 4 bottom left electrodes as cathodes and leave the other electrodes turned off (Fig. 12). Thus, the multi-electrode

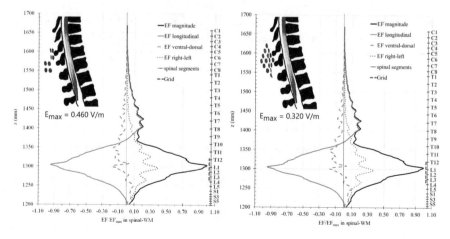

Fig. 12 Distribution of total EF magnitude and EF components in the diagonal grid (left) and grid rotation 45° (right) montages, averaged over 1-mm thick slices along the SC length and normalized to maximum values in each case. The EF magnitude volume distribution over T6-Co segments in WM is represented at the top left corner of each plot, with the value of E_{max}

grid can be used for differential targeting, i.e. to stimulate different spinal networks according to their orientation at specific periods within the same session, which can be of great relevance for clinical applications of tsDCS for multi-factorial spinal dysfunctions.

4.3   Limitations in Modelling tsDCS: What Lies Ahead

The present study only considers isotropic electrical conductivities. In a previous study, we included artificial anisotropy tensors of WM and muscle, considering the typical direction of fibers. Muscle anisotropy only contributed to change slightly the EF magnitude, however the anisotropy of WM increased the EF sensitivity to anatomical characteristics and spinal curvature, which can have an impact on the predicted neuromodulation on certain spinal regions. Thus, we recommend that the use of models to optimize tsDCS protocols should include, whenever possible, anisotropic considerations for the WM and GM.

The model presented here has an artificial design of the GM and does not represent the spinal rootlets with dorsal ganglia, where sensory neurons somas are located. Realistic representation of these components will be a difficult step, since it will require MRIs of very high resolution, with a segmentation procedure which will take many semi-automatic or even manual time-consuming tasks. Even so, this will be a necessary update of human realistic models to address the true neuromodulatory potential of tsDCS.

Connecting macroscale representations of the SC with neuronal and circuit models is also essential to determine what are the effects at neuronal and synaptic level. The long-term effects of tsDCS may be similar to the processes of long-term potentiation (LTP) observed in cortical stimulation, related with neuroplasticity [18, 25]. Neuronal and network models can unveil how tsDCS may be applied to repair spinal network communication and recovery of function at long-term, thus providing a solid support of tsDCS as an effective therapeutic resource for sensorimotor rehabilitation.

5   Conclusion

Non-invasive stimulation of the spinal cord in the form of tsDCS has a promising therapeutic potential to address sensorimotor dysfunctions of the spinal cord. Computational numerical studies can provide useful information to optimize the delivery of currents, indicating how to combine knowledge of the electrical properties of tissues and relative placement of electrodes to increase stimulation selectivity aiming at a specific spinal target. Even if not very different from traditional 2-electrode montages, multi-electrode grid stimulation paradigms can provide differential stimulation, by changing EF directions over different periods without changing the position of the electrode grid. This could be helpful to modulate different spinal circuits in a single session, when these vary in orientation inside the spinal cord. Furthermore, the grid system may be further optimized by adjusting electrode size and inter-electrode distance, since these characteristics can change EF magnitudes, as predicted in 2-electrode montages.

Acknowledgments Research supported by Fundação para a Ciência e Tecnologia (FCT) in the scope of the FCT-IBEB Strategic Project UIDB/00645/2020. M. Pereira was supported by a FCT grant, reference PD/BD/150340/2019.

References

1. F. Cogiamanian, G. Ardolino, M. Vergari, R. Ferrucci, M. Ciocca, E. Scelzo, et al., Transcutaneous spinal direct current stimulation. Front. Psych. **3**, 63 (2012)
2. F. Cogiamanian, M. Vergari, F. Pulecchi, S. Marceglia, A. Priori, Effect of spinal transcutaneous direct current stimulation on somatosensory evoked potentials in humans. Clin. Neurophysiol. **119**, 2636–2640 (2008)
3. T. Winkler, P. Hering, A. Straube, Spinal DC stimulation in humans modulates post-activation depression of the H-reflex depending on current polarity. Clin. Neurophysiol. **121**(6), 957–961 (2010)
4. M. Hubli, V. Dietz, M. Schrafl-Altermatt, M. Bolliger, Modulation of spinal neuronal excitability by spinal direct currents and locomotion after spinal cord injury. Clin. Neurophysiol. **124**(suppl 6), 1187–1195 (2013)
5. T. Bocci, B. Vanninia, A. Torzini, A. Mazzatenta, M. Vergari, F. Cogiamanian, et al., Cathodal transcutaneous spinal direct current stimulation (tsDCS) improves motor unit recruitment in healthy subjects. Neurosci. Lett. **578**, 75–79 (2014)
6. M. Parazzini, S. Fiocchi, I. Liorni, E. Rossi, F. Cogiamanian, M. Vergari, et al., Modelling the current density generated by transcutaneous spinal direct current stimulation (tsDCS). Clin. Neurophysiol. **125**(suppl. 11), 2260–2270 (2014)
7. S. Fiocchi, P. Ravazzani, A. Priori, M. Parazzini, Cerebellar and spinal direct current stimulation in children: Computational modeling of the induced electric field. Front. Hum. Neurosci. **10**, 522 (2016)
8. D.S. Kuck, E. van Asseldonk, Modeling trans-spinal direct current stimulation for the modulation of the lumbar spinal motor pathways. J. Neural Eng. **4**(suppl. 5), 056014 (2017)
9. S.R. Fernandes, R. Salvador, C. Wenger, M. de Carvalho, P.C. Miranda, Transcutaneous spinal direct current stimulation of the lumbar and sacral spinal cord: a modelling study. J. Neural Eng. **15**(suppl 3), 036008 (2018)
10. A. Datta, V. Bansal, J. Diaz, J. Patel, D. Reato, M. Bikson, Gyri-precise head model of transcranial direct current stimulation: improved spatial focality using a ring electrode versus conventional rectangular pad. Brain Stimul. **2**, 201–207 (2009)
11. J.P. Dmochowski, A. Datta, M. Bikson, Y. Su, L.C. Parra, Optimized multi-electrode stimulation increases focality and intensity at target. J. Neural Eng. **8**, 046011 (2011)
12. G.B. Saturnino, A. Antunes, A. Thielscher, On the importance of electrode parameters for shaping electric field patterns generated by tDCS. NeuroImage **120**, 25–35 (2015)
13. R.J. Sadleir, T.D. Vannorsdall, D.J. Schretlen, B. Gordon, Transcranial direct current stimulation (tDCS) in a realistic head model. NeuroImage **51**, 1310–1318 (2010)
14. L. Santos, M. Martinho, R. Salvador, C. Wenger, S.R. Fernandes, O. Ripolles, G. Ruffini, P.C. Miranda, Evaluation of the electric field in the brain during transcranial direct cur-

rent stimulation: a sensitivity analysis. Ann. Int. Conf. IEEE Eng. Med. Biol. Soc. **2016**, 1778–1781 (2016)

15. A. Christ, W. Kainz, E.G. Hahn, K. Honegger, M. Zefferer, E. Neufeld, et al., The virtual family-development of surface-based anatomical models of two adults and two children for dosimetric simulations. Phys. Med. Biol. **55**(suppl 2), N23–N38 (2010)

16. K. Zhu, L. Li, X. Wei, X. Sui, A 3D computational model of transcutaneous electrical nerve stimulation for estimating abeta tactile nerve fiber excitability. Front. Neurosc. **11**, 250 (2017)

17. P.C. Miranda, A. Mekonnen, R. Salvador, G. Ruffini, The electric field in the cortex during transcranial current stimulation. NeuroImage **70**, 48–58 (2013)

18. M.A. Nitsche, W. Paulus, Excitability changes induced in the human motor cortex by weak transcranial direct current stimulation. J. Physiol. **527**(3), 633–639 (2000)

19. D. Liebetanz, R. Koch, S. Mayenfels, F. Konig, W. Paulus, M.A. Nitsche, Safety limits of cathodal transcranial direct current stimulation in rats. Clin. Neurophysiol. **120**(1), 1161–1167 (2009)

20. M. Niérat, T. Similowski, J. Lamy, Does trans-spinal direct current stimulation Alter phrenic motoneurons and respiratory neuromechanical outputs in humans? A double-blind, sham-controlled, randomized, crossover study. J. Neurosci. **34**(43), 14420–14429 (2014)

21. N. Khadka, H. Borges, B. Paneri, T. Kaufman, E. Nassis, et al., Adaptive current tDCS up to 4 mA. Brain Stimul. **13**(1), 69–79 (2020)

22. A. Thielscher, A. Opitz, M. Windhoff, Impact of the gyral geometry on the electric field induced by transcranial magnetic stimulation. NeuroImage **54**, 234–243 (2011)

23. A. Caetano, M. Pereira, M. de Carvalho, A 15-minute session of direct current stimulation does not produce lasting changes in axonal excitability. Neurophysiol. Clin. **49**(4), 277–282 (2019)

24. A. Caetano, P. Pereira, M. Pereira, M. de Carvalho, Modulation of sensory nerve fiber excitability by transcutaneous cathodal direct current stimulation. Neurophysiol. Clin. **49**(5), 385–390 (2019)

25. Z. Ahmed, Electrophysiological characterization of spino-sciatic and cortico-sciatic associative plasticity: modulation by trans-spinal direct current and effects on recovery after spinal cord injury in mice. J. Neurosci. **33**(11), 4935–4946 (2013)

Part V
High Frequency Electromagnetic Modeling and Experiment: MRI Safety with Active and Passive Implants

RF-induced Heating Near Active Implanted Medical Devices in MRI: Impact of Tissue Simulating Medium

James E. Brown (✉), Paul J. Stadnik, Jeffrey A. Von Arx, and Dirk Muessig

1   Introduction

Generally, patients with active implanted medical devices (AIMD) are denied access to MRI due to the potential for hazardous interactions. It has been estimated that 17% of pacemaker patients will need an MRI within 12 months of device implantation [1]. MR imaging is a highly desirable imaging modality due to its non-ionizing radiation and image quality for soft tissue imaging [2]. Therefore, in recent years device manufacturers, academics, and regulatory agencies have developed standardized test methods for improving the access to this important diagnostic tool for AIMD patients [3, 4].

Several potentially hazardous interactions of implanted medical devices with the MR system have been identified [5, 6]. Computational human models (CHMs) are used in the evaluation of three hazards: RF-induced heating, RF-induced unintended stimulation, and RF-induced malfunction.

2   Utilizing Computational Human Models for the Assessment RF-induced Heating

Computational human models have been used for a range of electromagnetic applications [7]. This chapter discusses the specific application of CHMs to MRI RF-induced heating near AIMDs. While some studies have used measurement methods in homogenous phantoms [8], this work focuses on the methods outlined in [3, 4], which are based on the established transfer function method [9, 10]. The

J. E. Brown (✉) · P. J. Stadnik · J. A. Von Arx · D. Muessig
Micro Systems Engineering, Inc., Lake Oswego, OR, USA
e-mail: james.brown@biotronik.com

© The Author(s) 2023
S. Makarov et al. (eds.), *Brain and Human Body Modelling 2021*,
https://doi.org/10.1007/978-3-031-15451-5_8

benefits of using CHMs in this work have been outlined in [11, 12], which provide a framework for evaluating millions of scenarios and can be contrasted with the testing of pacemaker resilience to Electronic Article Surveillance (EAS) systems [13], which involves physically testing the device in a more limited number of prescribed tests. The general process is illustrated in Fig. 1.

The standardized test methods of [3] outline a 4-tier system for estimating *in vivo* RF-induced heating utilizing CHMs. Lower-numbered tiers are more conservative than higher-numbered tiers, that is Tier 1 is the most conservative and Tier 4 is considered the most accurate. Tier 1 is impractical and will be removed in the upcoming standard [14].

Tiers 2 and 3 both involve simulating a set of CHMs and a set of MR birdcage coils to build a library of expected field distributions within the human body. Examples of CHMs include the Virtual Family [15] and the Visible Human Project [16]. By varying parameters such as landmark position (alignment of the center of the MR coil with the CHM), body position, coil geometry, and using a variety of CHMs (different heights, weights, BMI, ages, *etc.*), millions of scenarios can be investigated. In a Tier 2 analysis, the risk to the patient is assessed using the maximum fields over the expected implant volume, after taking a 10 g average. In a laboratory setting, the device is then exposed to the appropriate field level and the temperature rise is measured. For elongated devices such as leaded AIMDs, as the maximum field values are taken over a large volume and then applied uniformly

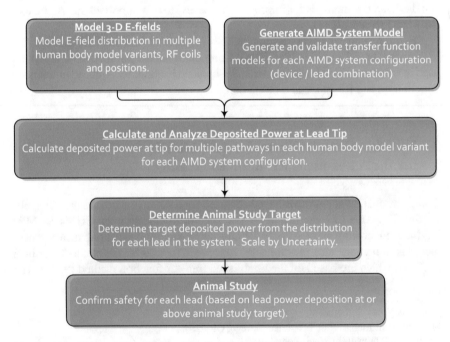

Fig. 1 Using CHMs to assess risk to the patient due to RF-induced tissue heating

over the entire device, this would result in impractically high test conditions and severely over-predict RF-induced heating. A Tier 3 analysis uses the same library of 3-D field distributions, but extracts the incident fields over one or more lead pathways and then combines this with the separately-derived lead model. Then, a probability distribution function of the estimated *in vivo* RF-induced heating can be carried forward to a safety assessment using an animal model or alternate method of clinical assessment.

The lead models used in Tier 3 analyses are often piecewise excitation models, where the lead response (*e.g.*, RF-induced deposited power or temperature rise near the lead tip) is quantified for the excitation of a section of the lead (*e.g.*, 1 cm). The so-called reciprocity method [10] involves a single excitation and a swept measurement of the response. Whether numerically- or experimentally-derived [17–21], transfer function models are developed within a homogenous medium, commonly referred to as a tissue-simulating medium (TSM). This work focuses on the application of the homogenous TSM used in the lead model to the inhomogeneous CHM used in the generation of the incident fields, and the ultimate impact to the *in vivo* estimation of RF-induced heating.

3 Tissue-Simulating Media (TSM)

Equivalent medium theorem [22] suggests that there exists an optimal homogenous TSM for agreement between the lead model (transfer function) and predicted *in vivo* RF-induced heating (in inhomogeneous CHMs). Indeed, recommended procedures for identifying such a medium are becoming standard practice [13] for AIMD manufacturers. The following subsections will discuss what is known about the impact of TSM on the AIMD model.

3.1 Effect of TSM on Computation of RF-induced Heating

The analytical problem of a medical device lead exposed to RF fields from an MR scanner is a particular instance of a wire scatterer in an incident field. Classically, these problems are investigated via the Method of Moments (MoM) [23], though MoM is known to be disadvantageous for analysing problems with multiple material boundaries as are present in the human body. Further, the problem of bare and insulated wire (as antennas or scatterers) in an extended conductive medium has been extensively studied in the literature [24–28], especially for submarine communication and geophysical exploration.

This analytical framework has been extended to the particular problem of implanted medical device leads [29], whereby the particular impact of the TSM to the predicted *in vivo* RF-induced heating can be quantified. Using a safety index which is proportional to the elevated SAR near the lead tip, general trends are seen for the impact of the TSM on resonant characteristics of the lead geometry. The resonant effect has also been extensively defined in the literature [30–32]. One particularly important consequence to the assessment of RF-induced heating near the lead tip is the necessity of elevated Electric field at the lead tip due to the discontinuity in the current distribution, which can be shown using a basic MoM formulation [33]. While the preceding analytical techniques focus on quantifying RF-induced heating in terms of electromagnetic quantities, additional considerations of the TSM for measuring temperature rise are also required [34, 35].

While these studies provide an understanding of the effect of changing the surrounding medium, or TSM, on the lead response, more contemporary research has focused on the impact of the choice of TSM on the overall accuracy of the *in vivo* prediction of RF-induced heating. This important work is discussed in the next two subsections, which are divided into studies of simplified structures such as a simple wire and more complex geometries representing actual geometries of practical AIMD leads.

3.2   Numerical Studies of Simplified Structures

A basic model for an AIMD lead is an insulated wire with a bare section at the distal end. This simplified structure is useful for its ease in construction and for simplifying analyses, and thus is used for standardization of test methods [3]. Indeed, such structures are often used for numerical simulations of transfer functions, examples include [18, 19].

A transmission-line model of a simplified structure was shown in [36] to give good agreement with measurement for a variety of parameters of the wire (insulation thickness, conductor diameter, insulator dielectric constant, lead length, *etc.*) and for different TSMs, for RF-induced voltage at the proximal end. While this work did not focus on RF-induced heating, it is still important for any discussion of the impact of TSM properties to highlight the influence of the tip impedance, Z_{tip}, on the ultimate result, a relationship which has been explored in further depth [37]. Still, neither of these works investigated which TSM would then give the most accurate result when compared to a full evaluation with an inhomogeneous CHM. This question for simplified structures is answered in [38], which concludes that the appropriate choice in homogeneous TSM is the TSM with the conductivity which matches that at the lead tip. This conclusion is dependent on the thickness of the insulation surrounding the conductor, as [39] shows that for a given TSM the SAR will plateau for increased insulation thickness. For thin insulation leads, the TSM at the lead tip is no longer dominant and the choice in TSM should align with the average along the lead body [40]. While the focus of this work has been RF-induced heating, most of these principles are extensible to other RF-induced hazards, such as RF-induced unintended stimulation [41].

3.3   *Numerical Studies of Realistic Device Geometries*

Medical device leads are of course more complex than the simplified structures of the previous subsection. For example, pacemaker leads are helically-wound, with multiple conductors present which may be coaxial or coradial [42]. The current version of the ISO technical specification [3] for MRI safety of AIMDs includes a set of suggested TSMs for manufacturers to consider. For electrically short, non-leaded devices the analysis may be more straightforward. An example of this analysis is shown in [43], which follows the test methods defined in [4]. In this type of analysis, typically two TSMs are used, and the worst-case of these two scenarios is carried forward. In some situations, this may be overly conservative.

Importantly, the set of geometries in [44] included up to four helically wound conductors and includes a parametric study of the helix design, to explore the relationship of the tip impedance to the predicted RF-induced heating. The importance of the lead pathway to the optimal choice of TSM is discussed, and a novel method for constructing an inhomogeneous computational phantom for transfer function generation is proposed.

For a percutaneous SCS lead, the choice of the high conductivity medium (HCM) from [3] was shown to be more accurate, while the low conductivity medium (LCM) would lead to an over-prediction of RF-induced heating by 72–74% [45]. This contrasts with another study of a simplified pacemaker lead [46] which shows that LCM is more accurate while the HCM overpredicted RF-induced heating by 80%. However, exact insulation thicknesses for the lead models used in these two studies were not given. Both studies illustrate the importance of TSM selection on the predicted risk to the patient due to RF-induced heating. While safety assessments should always be conservative, care must be taken to avoid situations where the prediction is so high as to prevent patient access to this important diagnostic tool.

4   Discussion

In this chapter, we have investigated the impact of the CHM on the estimation of MRI RF-induced heating near the tips of AIMD leads through the lens of TSM selection during transfer function development. While general guidance for TSM selection is still being investigated, early evidence exists to recommend that device manufacturers use a TSM similar to that found at the distal end for leads with thick insulation, while for thin insulation, a homogenous TSM should approximate the average properties of tissues in contact with the lead body. The data is supported by experimental, theoretical, and computational analysis of simplified structures as well as practical lead geometries.

4.1 Future Work

The techniques shown in this work are currently being extended to provide general guidelines for the selection of TSM and its applicability to the CHM. AIMD manufacturers desire to have a more defined transition from "thick insulation" to "thin insulation", a definition which may vary due to the particular lead construction (straight versus helical wires, number of conductors, diameter of the conductors versus outer diameter of the lead, *etc.*). In addition, these types of sensitivity analyses are enabled mostly by numerical simulation, as to parameterize physical quantities could make prototyping and testing time-consuming and expensive.

The further refinement of numerically derived transfer functions can also enable more solutions for mitigating MRI RF-induced heating. Current practice is still to build prototypes and experimentally confirm the reduction in expected RF-induced heatingRF-induced heating [47, 48]. The investigation of solutions can only be improved by a more thorough understanding of the influence of the computational model on the ultimate results of the safety assessment.

References

1. R. Kalin, M.S. Stanton, Current clinical issues for MRI scanning of pacemaker and defibrillator patients. Pacing Clin. Electrophysiol. **28**(4), 326–328 (2005)
2. M.S. Brown, R.C. Semelka, Concept of magnetic resonance, in *MRI: Basic Principles and Applications*, 3rd edn., (John Wiley & Sons, Hoboken, 2004), pp. 11–20
3. ISO/TS 10974:2018 (E), "Assessment of the safety of magnetic resonance imaging for patients with an active implantable medical device", 2018
4. ANSI/AAMI PC76:2021, Requirements and Test Protocols for Safety of Patients with Pacemakers and ICDs Exposed to MRI, 2021
5. L.P. Panych, B. Madore, The physices of MRI safety. J. Magn. Reson. Imaging **47**, 28–43 (2018)
6. J. Kabil et al., A review of numerical simulation and analytical modeling for medical devices safety in MRI. Yearb. Med. Inform., 152–158 (2016)
7. S.N. Makarov et al., Virtual human models for electromagnetic studies and their applications. IEEE Rev. Biomed. Eng. **10**, 95–121 (2017)
8. P. Nordbeck et al., Spatial distribution of RF-induced E-fields and implant heating in MRI. Magn. Reson. Med. **60**, 312–319 (2008)
9. S.-M. Park, R. Kamondetdacha, J.A. Nyenhuis, Calculation of MRI-induced heating of an implanted medical Lead wire with an electric field transfer function. J. Magn. Reson. Imag. **26**, 1278–1285 (2007)
10. S. Feng et al., A technique to evaluate MRI-induced electric fields at the ends of practical implanted Lead. IEEE Trans. Microw. Theory Techn. **63**(1), 305–313 (2015)
11. J.E. Brown, et al., MR Conditional Safety Assessment of Implanted Medical Devices: Advantages of Computational Human Phantoms, *Proc. 38th Annu. Int. Conf. IEEE EMBC*, Orlando, FL, pp. 6465–6468, 2016
12. B.L. Wilkoff et al., Safe magnetic resonance imaging scanning of patients with cardiac rhythm devices: a role for computer modeling. Heart Rhythm. **10**(12), 1815–1821 (2013)
13. R. Herkert, *E3 Test Protocol for Medical Devices to Security and Logistical Systems, Ver. 6.0* (Medical Device Test Center, Georgia Tech Research Institute, Atlanta, 2013)

14. ISO 10974 (Draft), Assessment of the safety of magnetic resonance imaging for patients with an active implantable medical device, To be published

15. A. Christ et al., The virtual family—development of surface-based anatomical models of two adults and two children for dosimetric simulations. Phys. Med. Bio. **55**, N23–N38 (2010)

16. G.M. Noetscher, et al., Computational Human Model VHP-Female Derived from Datasets of the National Library of Medicine, *Proc. 38th Annu. Int. Conf. IEEE EMBC*, Orlando, FL, 2016, pp. 3350–3353

17. A. Yao et al., Efficient and reliable assessment of the maximum local tissue temperature increase at the electrodes of medical implants under MRI exposure. Bioelectromagnetics **40**(6), 422–433 (2019)

18. M. Kozlov, W. Kainz, Comparison of lead electromagnetic model and 3D EM results for helix and straight leads, *Proc. 19th Int. Conf. Electromagn. Adv. Appl.*, pp. 649–652, 2017

19. M. Kozlov, W. Kainz, Lead electromagnetic model to evaluate RF-induced heating of a coax lead: a numerical case study at 128 MHz. IEEE J. Electromagn. RF Microw. Med. Biol. **2**(4), 286–293 (2018)

20. E. Zastrow, M. Capstick, N. Kuster, Experimental system for RF-heating characterization of medical implants during MRI", Proc. 24th Annu. Meeting ISMRM, Singapore, 2016

21. E. Zastrow, A. Yao, N. Kuster, Practical considerations in experimental evaluations of RF-induced heating of leaded implants, 32nd URSI GASS, Montreal, Canada, 2017

22. Y. Wang et al., On the development of equivalent medium for active implantable device radio-frequency safety assessment. Magn. Reson. Med. **82**, 1164–1176 (2019)

23. R.F. Harrington, *Field Computation by Moment Methods* (IEEE Press, New York, 1993)

24. P.E. Atlamazoglou, N.K. Uzunoglu, A Galerkin moment method for the analysis of an insulated antenna in dissipative dielectric medium. IEEE Trans. Microw. Theory **46**, 988–996 (1998)

25. R.W.P. King, G.S. Smith, *Antennas in Matter: Fundamentals, Theory, and Applications* (The MIT Press, Cambridge, MA, 1981)

26. R.W.P. King, B.S. Trembly, J.S. Strohbehn, The electromagnetic field of an insulated antenna in a conducting or dielectric medium. IEEE Trans. Microw. Theory Techn. **MTT-31**(7), 574–583 (1983)

27. R.W.P. King, Antennas in material media near boundaries with application to communication and geophysical exploration, part I: the bare metal dipole. IEEE Trans. Ant. Propagat. **AP-34**(4), 483–489 (1986)

28. R.W.P. King, Antennas in material media near boundaries with application to communication and geophysical exploration, part II: the terminated insulated antenna. IEEE Trans. Ant. Propagat. **AP-34**(4), 490–496 (1986)

29. C.J. Yeung, R.C. Susil, E. Atalar, RF safety of wires in interventional MRI: using a safety index. Magn. Reson. Med. **47**, 187–193 (2002)

30. J.E. Brown, C.S. Lee, Radiofrequency resonance heating near medical devices in magnetic resonance imaging. Microwave Opt. Technol. Lett. **55**(2), 299–302 (2013)

31. S.O. McCabe, J.B. Scott, Cause and amelioration of MRI-induced heating through medical implant lead wires, *21st Elect New Zealand Conference*, Hamilton, New Zealand, Nov 2014

32. S.O. McCabe, J.B. Scott, Technique to Assess the Compatibility of Medical Implants to the RF Field in MRI, Asia-Pacific Microwave Conference 2015 6–9 Dec 2015

33. J.E. Brown, Radiofrequency heating near medical devices in magnetic resonance imaging, Ph.D. dissertation, Bobby B. Lyle School of Engineering, Southern Methodist University, Dallas, TX, 2012

34. S.M. Park et al., Gelled vs. non-gelled phantom material for measurement ofMRI-induced temperature increases with bioimplants. IEEE Trans. Magn. **39**(5), 3367–3369 (2003)

35. C.D. Smith, J.A. Nyenhuis, K.S. Foster, A comparison of phantom materials used in evaluation of radiofrequency heating of implanted medical devices during MRI, *Proc. 23rd Annu. Int. Conf. IEEE EMBC*, Istanbul, Turkey, pp. 2311–2314, 2001

36. J. Liu et al., A transmission line model for the evaluation of MRI RF induced fields on active implantable medical devices. IEEE Trans. Microw. Theory Techn. **66**(9), 4271–4281 (2018)

37. J. Liu, et al., On the relationship between impedances of active implantable medical devices and device safety under MRI RF emission, *IEEE Trans. EMC*, 2019 (Early Access)
38. K.N. Kurpad, et al., MRI RF safety of Active Implantable Medical Devices (AIMDs): numerical study of the effect of conductivity of tissue simulating media on device model accuracy, *Proc. 26th Annu. Meeting Int. Soc. of Magn. Reson. Med.*, Paris, France, pp. 4075, 2018
39. P.A. Bottomley et al., Designing passive MRI-safe implantable conducting leads with electrodes. Med. Phys. **37**(7), 3828–3843 (2010)
40. J.E. Brown, et al., MRI safety of active implantable medical devices: numerical study of the effect of lead insulation thickness on the RF-induced tissue heating at the lead electrode, *43rd Annu. Int. Conf. IEEE EMBC*, 2021
41. J.E. Brown et al., RF-induced unintended stimulation for implantable medical devices in MRI, in *Brain and Human Body Modeling 2020: Computational Human Models Presented at EMBC 2019 and the BRAIN Initiative® 2019 Meeting*, ed. by S. Makarov et al., (Springer Nature, Cham, Switzerland), pp. 283–292
42. C. Tang et al., Initial experience with a co-radial bipolar pacing lead. Pacing Clin. Electrophysiol **20**(7), 1800–1807 (1997)
43. J.E. Brown et al., Calculation of MRI RF-induced voltages for implanted medical devices using computational human models, in *Brain and Human Body Modeling: Computational Human Modeling at EMBC 2018*, ed. by S. Makarov et al., (Springer Nature, Cham, Switzerland), pp. 283–294
44. J. Liu et al., Investigations on tissue-simulating medium for MRI RF safety assessment for patients with active implantable medical devices. IEEE Trans. EMC **61**(4), 1091–1097 (2019)
45. X. Min, S. Sison, Transfer functions of a spinal cord stimulation systems in mixed media and homogeneous media for estimation of RF heating during MRI scans, *Proc. 40th Annu. Int. Conf. IEEE EMBC*, Honolulu, HI, 2018, pp. 2048–2051
46. X. Min, S. Sison, Impact of mixed media on transfer functions with a pacemaker system for estimation of RF heating during MRI scans. Comput. Cardiol. **44**, 1–4 (2017)
47. P. Nordbeck et al., Reducing RF-related heating of cardiac pacemaker leads in MRI: implementation and experimental verification of practical design changes. Magn. Reson. Med. **68**, 1963–1972 (2012)
48. P. Serano et al., Novel Brain stimulation technology provides compatibility with MRI. Sci. Rep. **5**, 9805 (2015)

Computational Tool Comprising Visible Human Project® Based Anatomical Female CAD Model and Ansys HFSS/Mechanical® FEM Software for Temperature Rise Prediction Near an Orthopedic Femoral Nail Implant During a 1.5 T MRI Scan

Gregory Noetscher, Peter Serano, Ara Nazarian, and Sergey Makarov

1 Introduction

This tool partially described in Refs. [1–4] is a non-clinical assessment model used to predict the RF power deposition induced temperature rise near orthopedic implants as a function of implant geometry, location, material, and scan time. It is using the complete *in silico* multiphysics MRI environment and the anatomical human CAD model. The validated context of use is limited to a mid-aged or elderly female subject of 50–70 years old with a higher obesity (BMI or body mass index of 30–36) scanned in a 1.5 T full-body circularly polarized cylindrical MRI birdcage coil with two ports at 64 MHz. It is also limited to one yet critical implant geometry: a long femoral nail (a nearly straight long metal rod) which is subject to most excessive heating during long scan times at 1.5 T.

This tool can augment the widely used ASTM F2182 standard that measures RF implant heating in a homogeneous gel phantom by providing extra safety margins caused by the influence of the realistic heterogeneous human body. It can help to

G. Noetscher (✉)
Worcester Polytechnic Institute, Worcester, MA, USA
e-mail: gregn@nevaem.com

P. Serano
Ansys, Inc., Canonsburg, PA, USA

A. Nazarian
Musculoskeletal Translational Innovation Initiative, Carl J. Shapiro Department of Orthopaedic Surgery, Beth Israel Deaconess Medical Center, Harvard Medical School, Boston, MA, USA

S. Makarov
ECE Department, Worcester Polytechnic Institute, Worcester, MA, USA

identify the appropriate worst-case implant size, configuration, and orientation as a function of the scan protocol and required scan time (the output RF power). The tool does provide the estimates of SAR values in the regions around the implant. The tool does not provide heating or SAR estimates for gradient-coil induced power deposition for orthopedic implants.

A number of modeling and physical processes were tested to validate this tool. They include: (i) topological validation of the entire female CAD model (required approximately 4000 man hours for its construction); (ii) anatomical validation of the constructed human model by anatomical experts; (iii) validation of the human model and FEM software compatibility; (iv) SAR deposition validation in the 1.5 T full-body MRI birdcage coil (accuracy of 10% or better); (v) temperature rise validation in a phantom with the actual long femoral nail implant (accuracy of 20% against the experiment) and; (v) temperature rise validation in the detailed human model with the same femoral nail implant located at approximately the same depth (accuracy of 25% or 2.4 °C on average with the standard deviation of or 0.2 °C against the experiments with the phantom).

In the last case, the tool testbed predicted the higher temperature rise (by approximately 25% or 2.4 °C higher on average) at the implant tips than the *in vitro* experiments with the simplified gel phantom. An additional validation of the MDDT was therefore made against *in vivo* measurements in living human arm which indicated the temperature deviation of the MDDT from the *in vivo* experiment of only 10%.

This tool addresses the growing number of patient MRI scans and expected prevalence of patients with implanted medical orthopedic devices specifically mid-aged and elderly women who are most affected by osteoporosis and the associated bone fracture. Newly developed implants, as well as legacy medical implants without MRI safety information, need to be evaluated for safety in the MRI environment.

The widely used ASTM F2182 standard allows measuring RF implant heating in a homogeneous gel phantom. However, the human body anatomy is different from this phantom. For safety and completeness, it may be therefore desired to additionally estimate the heating of the same implant in a realistic heterogeneous anatomical human model.

This MDDT tool may help to estimate the induced RF heating of a metallic orthopedic implant in the detailed model of an elderly woman aged ~60 at the frequency of 64 MHz (1.5 T MRI scan). This tool can also help to identify the appropriate worst-case implant size, configuration, orientation, and an allowable scan protocol (coil power) by performing multiple simulations to determine the RF-induced temperature rise as a function of the scan protocol and the required scan time.

This tool has demonstrated that it accurately predicts absolute temperature rise for RF-induced heating with acknowledgement of the following limitations:

- Although all separate blocks of the modeling pipeline (human model, coil model, SAR values) were validated separately and in the general case, the end result – temperature distribution along an implant and as a function of time – is limited to one yet most critical implant geometry: a long femoral nail (a nearly straight metal rod) subject to the most excessive heating and possible resonant effects.

- The focus of this tool is currently at 1.5 T/64 MHz cylindrical bore MRI systems using circularly polarized RF birdcage full body coils.
- The tool currently does not include multi-parts orthopedic implants or implants with screws.
- The tool currently does not provide heating estimates for gradient-coil induced power deposition for orthopedic implants.
- The computational human model margins are limited to women aged 50–70 and with a high obesity (BMI of 30–36).

The main advantage of using this tool lies in the possibility to accurately estimate temperature rise near an orthopedic implant in the realistic high-resolution elderly female human model and thus additionally justify and assess measurements obtained with the ASTM F2182 standard and their safety margins.

While modeling SAR distributions near implants in realistic virtual human models is well understood, accurate modeling and prediction of temperature rise at and near implants in human models is much more difficult. The reason is a necessity to couple an electromagnetic software and a thermal software with a realistic heterogeneous human model including fine implant and tissue details which are best described with the finite element method that is developed for curved and fine geometries.

The present tool addresses this knowledge gap by constructing and validating the state-of-the-art multiphysics FEM modeling pipeline. It couples the detailed accurate CAD human model, the world-best Ansys electromagnetic modeling software HFSS including the MRI coil modeling, and the Ansys thermal software, all within one single user-friendly shell – the Ansys Workbench.

The second advantage of using this tool is the type of the embedded anatomical female CAD model that is appropriate for elderly female subjects of 50–70 years old with a high obesity. There is no similar anatomical CAD model currently available.

The main disadvantage with the tool is that it currently requires long computation times necessary to obtain accurate and reliable temperature rise results. These times are on the order of 6–24 hours depending on the computer hardware used.

As stated previously, the tool's context of use is currently restricted to the long femoral nails and does not include other types of the femoral and shoulder orthopedic implants. It is also is limited to the 1.5 T cylindrical bore MRI system only.

2   Tool Description and Validation

2.1   VHP Female v 5.0 Anatomical Human Model (DOI 10.20298/VHP-FEMALE-V.5.0)

The VHP Female v.3.0–5.0 computational human model (also known under the nickname 'Nelly' to users of Dassault Systèmes SIMULIA software CST), shown at left in Fig. 1, is an anatomically accurate heterogeneous female (~60 years old, ~88 kg, BMI of ~36, obese, with heart pathology) surface-based human body model. It was constructed from the photographic cryosection data of the Visible Human Project of the US National Library of Medicine with the world-best isotropic resolution of 0.33 mm. Its construction has been well documented in the literature [5–10], but several points are worth emphasizing here:

1. There are 249 distinct components or triangular 2-manifold surface meshes (with an additional 40 characterizing embedded implants). No intersections or joint faces between discrete meshes are allowed, enabling unique assignment of electromagnetic, thermal, or other material properties.
2. The source data for the model are freely and publicly available [11, 12]. The complete co-registration data for all model cross-sections are also made publicly available [13].

One advantage of the model is its topological and computational simplicity. The model size is purposely limited to approximately 0.4 M triangular facets in total with an attempt to keep the anatomical accuracy within the body within 1–7 mm. This makes it possible to apply virtually any commercial or custom-tailored finite-element or boundary element computational solver in a very reasonable amount of time. Nearly identical results are obtained when using different software packages and numerical methods [14]. Yet another advantage is the separation of the muscular system of the body into individual muscle objects (about 50 in total).

Some recent uses of the VHP Female v3.0–5.0 computational human model have been documented in the literature – cf. for example [14–31] – and compared with experimental data on radio frequency propagation. A number of studies have focused on the characteristics and behaviors of antennas near the human body [15–23]. Others have examined through simulation various biomedical applications including transmission channel modeling [24], transcranial direct current stimulation [25], estimation of bone density [26], gastroenterology [27, 28], *SAR* simulations in MRI coils at 3 T [29], and safety of active implants under MRI procedures [30, 31].

The VHP-Female model has been licensed by several private entities engaged in medical device modeling including Ansys, Inc., Dassault Systèmes SIMULIA, Stryker Corp., Cambridge Consultants, Inc., and WIPL-D. A simplified version of the model, VHP-Female v. 2.2, has over 500 registered college users and is available for free download online.

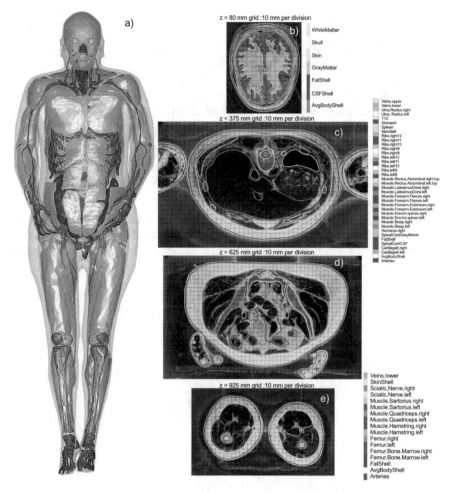

Fig. 1 (**a**) Full-body Visible Human Project Female v5.0 CAD based computational phantom embedded into Ansys FEM software. The phantom is composed of 249 individual structures. Some individual muscles are removed for clarity. (**b–e**) Examples of co-registration maps for surface meshes in different transverse planes with tissue labeling. In case (**b**), non-anatomical separation between scalp and skull was corrected. In case (**d**), the tissue labeling list is not shown. The complete full-body co-registration maps with the vertical resolution of 1 mm and with tissue labeling in every cross-sectional plane are available online in *.mp4 format [13] for independent inspection and verification purposes

2.2   Ansys FEM Computational Software Suite

2.2.1   Computation of SAR

The human model has been coupled with the proven Ansys Electronics Desktop (HFSS) finite-element electromagnetic software with adaptive mesh refinement. It has been employed to solve Maxwell's equations in three-dimensional space. In this way, the model is considered as an arbitrary (inhomogeneous) isotropic medium with piecewise-constant electric permittivity ε having the units of F/m and with constant magnetic permeability μ having the units of H/m.

After Maxwell's equations for an electric field (or the electric field intensity) $E(r, t)$ [V/m] and for a magnetic field (or the magnetic field intensity) $H(r, t)$ [A/m] are solved, the local SAR (W/kg) is defined through averaging the dissipated power per unit mass over a small (ideally infinitesimally small) volume V, that is

$$SAR(r) = \frac{1}{V} \int_V \frac{\sigma(r)}{2\rho(r)} |E(r)|^2 \, dV \tag{1}$$

where σ is (generally piece-wise constant) medium conductivity with the units of S/m, $\rho(r)$ is the local mass density, and $|E(r)|$ is the electric field magnitude at the observation point. The body-averaged or the whole-body SAR_{body} is given by averaging over the entire body volume, as

$$SAR_{body} = \frac{1}{V_{body}} \int_{V_{body}} \frac{\sigma(r)}{2\rho(r)} |E(r)|^2 \, dV \tag{2}$$

Similarly, SAR_{1g} is given by averaging over a volume with the weight of 1 g

$$SAR_{1g}(r) = \frac{1}{V_{1g}} \int_{V_{1g}} \frac{\sigma(r)}{2\rho(r)} |E(r)|^2 \, dV \tag{3}$$

$SAR_{10g}(r)$ is found in a similar fashion.

2.2.2 Computation of Temperature Rise

The SAR is the density of volumetric heat sources, not the temperature rise itself. To compute the resulting local temperature rise, the solutions of electromagnetic simulations have been coupled to Ansys Mechanical/Thermal via Ansys Workbench. Pennes' bioheat equation, based on the heat diffusion equation, is a standard approximation for heat transfer in biological tissues [39–41] implemented in the Ansys Mechanical software. It has the form

$$\rho(r)C(r)\frac{\partial T}{\partial t} = \nabla \cdot (k(r)\nabla T) = \rho(r)Q + \rho(r)SAR - \\ \rho_b C_b B(T - T_b) = 0 \tag{4}$$

where $T(r, t)$ is the local temperature, $\rho(r)$ is density, $C(r)$ is specific heat capacity (at constant pressure), Q is the metabolic heat generation rate, B is the perfusion rate [1/s]; index b is related to blood.

All electromagnetic, mechanical, and thermal tissue properties used in the present study are catalogued in the reputable IT'IS database [42]. The electromagnetic material properties are given as a function of frequency, which is ideal for the present analysis.

2.3   Validation of Overall Segmentation and Anatomical Correctness of the Human Model

2.3.1   Topological Validation

All 249 distinct components of the human model have been proven to be strictly 2-manifold (watertight) surface meshes or a reasonable mesh quality. All intersection or joint faces between discrete meshes have been eliminated enabling unique assignment of electromagnetic, thermal, or other material properties.

2.3.2   FEM Software Compatibility Validation

The entire model (including the implant parts) was so constructed that is passes the optional built-in Ansys mesh intersection and mesh quality checker with the highest option "Strict". Such a check can be independently performed and confirmed by any MDDT user.

2.3.3   Anatomical Validation

To validate the model anatomically, the co-registration method has been used, which implies direct superposition of transverse cross-sections of all surface tissue meshes onto the original cryosection images. An in-house MATLAB module (Fig. 1b–e) was written that performs such superposition with the resolution of 1 mm and simultaneously labels all tissue meshes which are present for a given cross-section. Its output is a scanning sequence in *.mp4 format [13].

After resolving multiple mesh intersections, the validation of the model segmentation shown in Fig. 1a was performed by

(i) co-registration of surface meshes superimposed onto the original cryo-section images (Fig. 1b–e) and;
(ii) correcting all intersection flaws.

Further visual anatomical validation was performed by a number of anatomical experts in their respective areas from Beth Israel Deaconess Med. Ctr. and

Massachusetts General Hospital, Boston MA including Profs A. Nazarian (ortho-
paedics), A. R. Opotowsky, (cardiovascular systems), V. Poylin, (gastroenterology),
E. K. Rodriguez (nusculoskeletal tissue components), A. Pascual-Leone (cranial
and intracranial anatomy) and Prof. G. Haleblian (urology). The questionable surface meshes (mostly bones but also soft tissues including scalp, bladder, uterus, etc.)
were corrected. An example of a non-anatomical flaw is shown in Fig. 1b where the
scalp was non-anatomically separated from the skull during the cryogenic process,
which required proper mesh adjustment.

The surface deviation for all meshes was found not to exceed 1–8 mm. The highest accuracy was achieved for

(i)   extracerebral and intracranial volume;
(ii) vertebral column/spinal cord.

The complete full-body co-registration maps (movies) with the vertical resolution
of 1 mm and with tissue labeling in every cross-sectional plane are available online
to the MDDT users in *.mp4 format [13] for independent inspection, verification
and design/development purposes.

2.4   Validation of Overall SAR Prediction by the MDDT During a 1.5 T MRI Scan

The first step of the MDDT workflow – the computation of local SAR – has been
validated by comparison of normalized SAR predicted by two different modeling
techniques in a 1.5 T birdcage whole body MRI coil.

A generic, whole-body high-pass birdcage coil with 16 rungs and 32 matching
capacitors, loaded with the VHP Female model, has been considered. The coil has a
diameter of 64 cm and length of 69 cm consistent with [46]. The simulation geometry is shown in Fig. 2a. Simulations have been conducted at shoulder/heart and
abdominal landmarks. For the former, the coil center is oriented to coincide with the
top of the T7 vertebra; the latter has the coil center located at the top of the L1 vertebra. The coil was tuned to the desired frequency of 64 MHz (B_0 = 1.5 T) when
loaded with the subject at each landmark (cf. [53]). Similar to [53], an ideal excitation was applied with 32 sources placed in the two end-rings to perform the function
of the capacitors. This excitation provides results which are very similar to the conventional two-port or four-port excitations [53].

A comparison was further made with the results of Ref. [53] which was using a
nearly identical high-pass birdcage coil (diameter of 63 cm and length of 70 cm),
the nearly identical heart landmark, and the identical coil excitation type. However,
an in-house voxel model for the same VHP-Female dataset was employed in [53]
followed by the FDTD simulation method with the resolution of 5 mm.

To compare the solution variation as a function of FEM mesh density and the
solution convergence trends, solutions were generated first with 1 and then with 8
adaptive mesh refinement passes, created approximately 0.5 M and 2.0 M tetrahedra,
respectively. As an example, Fig. 2 shows the corresponding local *SAR* distributions
at two different FEM resolutions in the coronal plane for the coil loaded with the

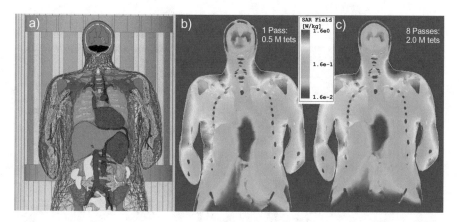

Fig. 2 Local SAR distribution in the coronal plane for a high pass full-body RF coil operating at 64 MHz loaded with the VHP-Female v5.0 computational phantom given $\boldsymbol{B}_1^+$ amplitude of 1 μT at the coil center. (**a**) Positioning of model within the birdcage at shoulder/heart landmark. (**b**) Ansys HFSS (Electronics Desktop) solution with one adaptive pass. (**c**) Ansys HFSS (Electronics Desktop) solution with eight adaptive passes

VHP-Female v3.0 computational human model given $\boldsymbol{B}_1^+$ amplitude of 1 μT at the coil center. Fields solutions from Ref. [53] were also normalized given the desired magnitude of $\boldsymbol{B}_1^+$ at the coil center of 1 μT. The normalization is done in the form

$$SAR \rightarrow \frac{SAR}{\left(\boldsymbol{B}_1^+ / 1\mu T\right)^2} \tag{6}$$

The local *SAR* was computed in Ansys HFSS and then exported to MATLAB over a uniform 3D grid of 2 mm in size. Whole-body SAR_{body} was computed from this data directly in MATLAB. SAR_{1g} was also calculated by finding a volume surrounding the observation point having the mass of exactly 1 g, and then performing averaging according to Eq. (3c). This averaging volume contains approximately $5 \times 5 \times 5$ individual voxels ($2 \times 2 \times 2$ mm each) closest to the observation point. The observation points form a 3D sub-grid spaced of 20 mm and 10 mm, respectively. The tissue density was set at 1 g/cm³ uniformly in space. SAR_{10g} was computed in the same way. In this case, the averaging volume contained approximately 1250 individual voxels ($2 \times 2 \times 2$ mm each) closest to the observation point.

Figures 2b, c show the local *SAR* within the various tissues of the VHP-Female v3.0 model after 1 pass and 8 passes, respectively, for the shoulder/heart landmark. All results are normalized to the $\boldsymbol{B}_1^+$ amplitude of 1 μT at the coil center. By analyzing Fig. 2 we can see that

1. The accurate solution with eight adaptive mesh refinement passes generates a more realistic *SAR* distribution, especially with regard to the local *SAR* – see Table 1. In particular, two non-physical maxima of the *SAR* observed at the top of the head are no longer present.

Table 1 Comparison of *SAR* values predicted using the VHP-Female surface CAD model with the values predicted by the reference voxel model derived from the identical image dataset [53]. Nearly identical high-pass birdcage coil dimensions, coil landmark, and excitation type were used. All results are normalized to 1 μT $\boldsymbol{B}_1^+$ field at the coil center

Source	Method	Model	Coil landmark	Whole-body *SAR*	Max. non-averaged local *SAR*	Max. 1 g local *SAR*	Max. 10 g local *SAR*
Present report	FEM – 1 adaptive pass (Ansys)	CAD VHP-fem. v. 3.0 88 kg 64 MHz	Shoulder/ heart (top of vert. T7)	0.16	44.5	5.22	2.61
Present report	FEM – 8 adaptive passes (Ansys)	CAD VHP-Fem. v. 3.0 88 kg 64 MHz	Shoulder/ heart (top of vert. T7)	0.13	12.0	1.61	1.37
Ref. [53]	FDTD Voxel size 5 mm	Voxel Vis. Human Female 64 MHz	Heart	0.12	NA	1.78	NA

2. The accurate solution with eight adaptive mesh refinement passes and the coarse solution with one adaptive pass generate approximately the same whole-body *SAR* and *SAR* distribution maps, but considerably different peak local *SAR* values – see Table 1. Note that the coil has been retuned separately in both cases.
3. The maximum *SAR* for the present landmark is observed in the upper shoulder/ neck area and in the arms area.
4. A gap between the arms and the body may generate large *SAR* values.

Table 1 compares the computed *SAR* values with the values obtained in Ref. [53] under nearly identical conditions. For both models and both methods, the whole body *SAR* and the maximum $SAR_{1g}(\boldsymbol{r})$ differ by 8% and 10%, respectively.

A comparison with a variety of other modeling reference sources [35–37, 46–56] has also been made both at 1.5 T and 3 T. It was found that the *SAR* results calculated when using the VHP Female v3.0 model are strictly within the bounds of all other reported values.

2.5 Validation of Temperature Rise Prediction by the MDDT Near a Long Femoral Nail Implant Against Measurements with an ASTM-Like Phantom

ASTM and ASTM-like phantoms have been used as the standard method for testing compliance of implants within MRI environment [54]. These phantoms are not an equivalent of intricate human body structures. Here, the comparison is made with the original experiments performed in a series of papers [43–45].

2.5.1 Modeling Testbed

Prior to conducting simulations using the VHP Female v5.0 model, a necessary workflow and RF power calibration have been established. Figure 3 depicts to scale simulation of the original experiment performed in a series of papers [32–34]. It was also used for initial calibration purposes. At left in Fig. 3, a homogenous experimental AGAR gel phantom with electromagnetic properties matching that of the human body is positioned within a 1.5 T birdcage MRI coil operating at 64 MHz.

This phantom (40 cm long by 20 cm wide) in Fig. 3 is not the exactly the ASTM F2182 phantom. However, it does support and even enhances the loop-like distribution of the electric field in the birdcage coil and thus describes a realistic heating scenario, perhaps even the worst-case scenario with respect to implant heating.

All examinations [32–34] were performed with a 1.5 T MR scanner (MAGNETOM Symphony, SIEMENS) with the phantom at the center of the coil. A birdcage shaped transmit/receive body coil was used there with the inner diameter of 60 cm and the length of 70 cm, which is close to the present modeling setup (64 cm and 69 cm, respectively). Coil tuning and excitation was performed as described in the previous example. We again use ideal excitation which is very similar [53] to the two-port excitation used in [32–34].

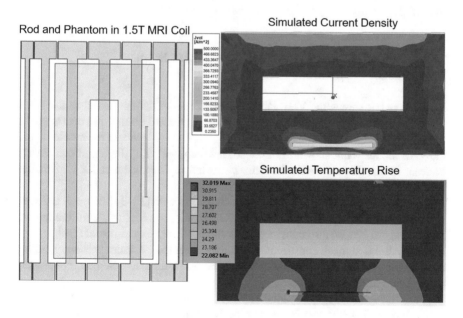

Fig. 3 Simulations to establish calibration for experiments conducted in [32–34]. At left, the metallic nail implant placed within a homogeneous loop-like Agar phantom [32–34] at a depth of 2 cm. At top right, the current density produced with a whole-phantom SAR of 4.0 W/kg. At bottom right, the simulated temperature given a total volumetric power loss of 120 W exactly corresponding to experiment [32–34]. Simulation results produced a temperature rise of 10.02 °C, slightly less than the 12.6 °C experimentally observed in [32–34]

A 24 cm long metallic orthopaedic nail implant of Zimmer, Inc. made of non-magnetic stainless still (originally very slightly bent but modeled as a straight rod of the same diameter) has been used. Such implants are normally used to treat fractures in clinical practice.

The implant was embedded in this phantom such that it is 2 cm away from the top and side edges of the phantom. Both the phantom and metallic rod have been assigned dimensions and material properties consistent with experiment [32–34].

2.5.2   Replication of Original Experiments

The computed volumetric current density within the phantom is shown at top right in Fig. 3. This density matches well with the published result; it was produced by adjusting coil power to exactly replicate the RF exposure given in [32–34] – a whole-phantom *SAR* of 4.0 W/kg and a volumetric power loss density of 120 W in the phantom. The temperature simulation is shown at bottom right of Fig. 3. Given the above power loss density as an internal heat generation source, the total temperature rise within the human model was 10.02 °C. This value is slightly less than the 12.6 °C observed in [32–34]. However, it is close enough (a 20% temperature difference) to give confidence in the adequate simulation setup, enabling the extension of this methodology to the case involving the VHP female human model.

2.5.3   Heating of the Same Implant in the MDDT Human
Female Phantom

Figure 4 at left shows the VHP Female computational human model in the 1.5 T birdcage MRI coil. The human model has been oriented so that the center of the femur bone is aligned with the center of the coil. The left quadriceps muscle within the VHP model is not shown in Fig. 4 so that the position of the femur can clearly be seen. Within the left femur, the same 24 cm long cylindrical metallic implant has been inserted as required in the surgical practice.

This rod is assigned material properties consistent with [32–34]. The coil is driven at 64 MHz and at a power such that the volumetric power loss within the human model is again 120 W, also consistent with the published experimental results.

A length-based mesh constraint of no edge larger than 2.5 mm was enforced for the metallic rod mesh and a total of about 636,000 tetrahedral elements were used in the Ansys HFSS simulation.

Once the electromagnetic simulation was complete, the results were passed to Ansys Mechanical software by linking the two simulations in Ansys Workbench. A mesh refinement was again employed on the faces of the rod to ensure that a dense enough mesh was created to capture the local temperature changes. Approximately 629,000 tetrahedral elements were used in the transient thermal simulation. Volumetric losses (*SAR*) produced by the Ansys HFSS simulation were imported into Ansys Mechanical and used as the internal heat generation source density. The RF coil and the

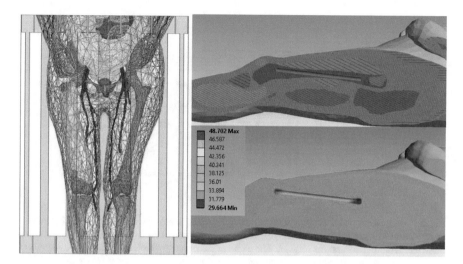

Fig. 4 The VHP-Female computational phantom positioned with the 1.5 T MRI birdcage coil. Some body parts are removed for clarity. At left, the femur position is illustrated to show its orientation within the model – the metallic nail implant is aligned to reside within the trabecular bone structure. At top right, each individual object within Ansys Mechanical model is assigned specific thermal properties. At bottom right, the temperature rise is shown after 900 seconds of continuous coil operation. These values correspond well with published experimental data [32–34]

heat sources were active for 900 s and the model was allowed to cool for another 600 s. The values for the implant temperature captured from 0 to 900 s are shown in Fig. 5.

Figure 5 shows temperature dynamics: it compares simulated numerical values (red stars) obtained using the VHP Female computational phantom with published experimental data (two black curves) [32–34] for the maximum temperature rise near the implant. The depth of the implant within the VHP model is *approximately* 3–4 cm and the maximum temperature value after 900 seconds of coil operation time is represented by the top red star.

Peak simulated temperature values for the implant within the human model are shown in the bottom right of Fig. 4. A maximum temperature rise of approximately 11 °C is computed at the very ends of the metallic implant against 8.25 °C observed in experiment at 4 cm depth [32–34]. Thus, the MDDT testbed predicts a higher (by 2.75 °C) maximum temperature rise of 8.25 °C than what was measured in [32–34] which constitutes the difference of 33% or 2.75 °C.

This is likely due to the different (non-homogeneous) material properties employed in the present study, the realistic angled orientation of the nail, and a possible resonant behavior of the long implant. In the VHP Female model, the rod is aligned with the femur and represents a more realistic position that would be encountered in a clinical setting.

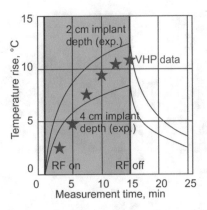

Fig. 5 Comparison of the simulated numerical values (red stars) obtained using the VHP Female computational phantom with published experimental data (two black curves) [32–34] for the maximum temperature rise near the implant. The depth of the implant within the VHP model is *approximately* 4 cm and the maximum temperature value after 900 seconds of coil operation time is represented by the top red star. The VHP model predicts a slightly higher (by 2.75°) maximum temperature rise than what was measured in [32–34]. This is likely due to the different (non-homogeneous) material properties employed in the present study and the slightly angled orientation of the nail

2.6 Variation and Uncertainty of Temperature Rise Measurements Near the Implant

Table 2 gives the least mean square difference in the temperature rise (maximum implant temperature at the tip is recorded) over the entire interval from 0 to 15 min between the modeled data and the in-vitro data at 2 and 4 cm implant depth, respectively, in degrees C. Statistical significance of the deviation is quantified via the p-value for the paired-sample t-test with $p \leq 0.05$ considered statistically significant.

2.7 Validation of Temperature Rise Prediction by the MDDT Against In Vivo Measurements

In the previous case validation example, the MDDT testbed predicted the higher maximum temperature rise (up to 33% higher) at the implant tips than the experiment *in vitro* with the simplified gel phantom. An additional validation of temperature rise was therefore made against *in vivo* measurements in living human.

Since *in-vivo* measurements in the MRI coil at 1.5 T have not been reported, a comparison was made with tissue heating due to a small single-loop coil close to skin surface at 165 MHz [38]. The replication of the experimental setup [38] is shown in Fig. 6. The forearm of the VHP Female computational human model has

Table 2 Least mean square difference in the temperature rise (maximum implant temperature at the tip) over the entire interval from 0 to 15 min between the modeled data and the *in-vitro* data at 2 and 4 cm depth in degrees C

Phantom vs. MDDT	Deviation: 2 cm implant depth vs. left leg	Deviation: 4 cm implant depth vs. left leg	Deviation: 2 cm implant depth vs. right leg	Deviation: 4 cm implant depth vs. right leg
	2.3 °C	2.5 °C	2.3 °C	2.5 °C

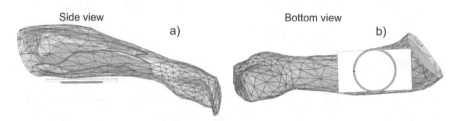

Fig. 6 The forearm of the VHP-Female computational human model adjacent to a single loop coil operating at 165 MHz in Ansys HFSS. The model geometry is shown, including internal forearm tissues, each defined with specific electromagnetic and thermal properties

been isolated to speed up the simulation. Internal structures, including the humerus, ulna and radius bones, extensor, flexor, triceps and biceps muscles, radial nerve, and various arteries and veins, are encapsulated within concentric layers of muscle, fat and skin tissue. The coil [38] is modeled as a 80 mm diameter copper torus with a minor diameter of 2 mm driven at 165 MHz by a 50 Ohm lumped antenna port and a thin sheet of Teflon separates the antenna from the forearm. These settings exactly correspond to experiment. All thermal and electromagnetic material properties associated with the internal body structures correspond to those supplied in [38].

According to [38], the coil antenna is provided sufficient power to dissipate approximately 31 W within the forearm and produce peak local *SAR* values in the lower corners of the forearm of approximately 450 W/kg. These power conditions have been replicated in simulations with the forearm of the VHP Female model. Adaptive mesh refinement produced a volumetric mesh consisting of approximately 0.1 M tetrahedra.

Following simulation in Ansys HFSS, the peak *SAR* value experienced in the outermost model layer is computed as approximately 450 W/kg, consistent with the value reported in [38] to within 5%.

Then, the geometry and resulting fields were passed via Ansys Workbench to Ansys Mechanical. Here, the human model was again remeshed, producing just over 0.06 M elements. The volumetric power losses determined in Ansys HFSS were imported and evaluated as the heat generation source density. Heat generation sources were active for 120 s and the model then cooled via convection for 19.2 s; temperatures were recorded throughout the duration of the simulation. The corresponding temperature rise after 139 seconds in total was computed and compared with experiment.

The simulated and measured [38] thermal maps within the forearm in a plane directly above the coil center have been examined. The peak simulated temperature change is about 6.6 °C, which is slightly higher that the measured 6 °C change reported in [38]. The deviation is within 10%.

The spatial temperature map are also consistent with the measured map despite the different wrist geometry and tissue composition. There is some difference in the observed depth of temperature change. This is likely mostly due to the fat layer of the VHP Female model which is thicker than the subject reported in [38].

3 Conclusions

Based on the evidence provided, this non-clinical assessment tool entitled "Computational Tool Comprising Visible Human Project® Based Anatomical Female CAD Model and Ansys HFSS/Mechanical® FEM Software for Temperature Rise Prediction near an Orthopedic Femoral Nail Implant during a 1.5 T MRI Scan" was found to reliably predict both temperature distribution and its evolution in time along the long femoral nail metal implant caused by RF power deposition from the 1.5 T birdcage MRI full body coil. The tool can also help to identify the appropriate worst-case device and coil size, configuration, and orientation by performing multiple simulations to determine the RF-induced temperature rise as a function of a scan protocol and required scan time.

All separate blocks of the modeling pipeline – the human model topology and anatomy including co-registration, surgically correct implant embedding, the RF coil model, and the resulting SAR and temperature behavior – have been validated independently and all the validation results have been made available to the user.

References

1. G.M. Noetscher, P. Serano, W.A. Wartman, K. Fujimoto, S.N. Makarov, Visible Human Project® female surface based computational phantom (Nelly) for radio-frequency safety evaluation in MRI coils. PLoS One 16(12), e0260922 (2021). https://doi.org/10.1371/journal.pone.0260922. PMID: 34890429; PMCID: PMC8664205
2. S.N. Makarov, G.M. Noetscher, J. Yanamadala, M.W. Piazza, S. Louie, A. Prokop, A. Nazarian, A. Nummenmaa, Virtual human models for electromagnetic studies and their applications. IEEE Rev. Biomed. Eng. 10, 95–121 (2017). https://doi.org/10.1109/RBME.2017.2722420. Epub 2017 Jun 30
3. H. Tankaria, X.J. Jackson, R. Borwankar, G.N. Srichandhru, A. Le Tran, J. Yanamadala, G.M. Noetscher, A. Nazarian, S. Louie, S.N. Makarov, VHP-female full-body human CAD model for cross-platform FEM simulations: recent development and validations. Annu. Int. Conf. IEEE Eng. Med. Biol. Soc. 2016, 2232–2235 (2016). https://doi.org/10.1109/EMBC.2016.7591173
4. M. Kozlov, G.M. Noetscher, A. Nazarian, S.N. Makarov, Comparative analysis of different hip implants within a realistic human model located inside a 1.5T MRI whole body RF coil.

Annu. Int. Conf. IEEE Eng. Med. Biol. Soc. **2015**, 7913–7916 (2015). https://doi.org/10.1109/EMBC.2015.7320227

5. G.M. Noetscher et al., VHP-Female v3.0 FEM/BEM Computational Human Phantom, *24th Int. Meshing Roundtable (IMR24)*, Austin, TX, Oct. 12–14, 2015

6. J. Yanamadala et al., Multi-purpose VHP-female version 3.0 cross-platform computational human model, *10th European Conf. on Antennas and Propagation (EuCAP16)*, Davos, Switzerland, April 2016, pp. 1–5

7. H. Tankaria et al., VHP-female full-body human CAD model for cross-platform FEM simulations – recent development and validations, *38th Annual Int. Conf. of the IEEE Engineering in Medicine and Biology Society (EMBC 2016)*, Orlando, FL, Aug. 16–20, 2016

8. G. Noetscher et al., Creating a computational human model, *IEEE Pulse*, April 27th 2016. Available: http://pulse.embs.org/march-2016/creating-a-computational-human-model/

9. G. Noetscher et al., Computational human model VHP-FEMALE derived from datasets of the national library of medicine, *38th Annual Int. Conf. of the IEEE Engineering in Medicine and Biology Society* (EMBC 2016), Orlando, FL, Aug. 16–20 2016

10. G.M. Noetscher et al., VHP-female CAD human model family for antenna modeling, *2016 IEEE Int. Sym. on Antennas and Propagation/USNC-URSI National Radio Science Meeting*, Puerto Rico, June 2016

11. The Visible Human Project, US National Library of medicine. Online: https://www.nlm.nih.gov/research/visible/visible_human.html

12. M.J. Ackerman, The visible human project: from body to bits. IEEE Pulse **8**(4), 39–41 (2017). https://doi.org/10.1109/MPUL.2017.2701221

13. VHP Female v.3.0–5.0 Co-Registration Standard. 08/23/2017. Online: https://www.dropbox.com/sh/f1unwyan9e0sh2v/AACXzcQ7EpR8wYXowbnhC-eAa?dl=0

14. J. Yanamadala, G. Noetscher, S. Louie, A. Prokop, M. Kozlov, A. Nazarian, S.N. Makarov, Multi-purpose VHP-female version 3.0 cross-platform computational human model, *2016 10th European Conference on Antennas and Propagation (EuCAP)*, Davos, Switzerland, 2016, pp. 1–5, https://doi.org/10.1109/EuCAP.2016.7481298

15. R. Lemdiasov, A. Venkatasubramanian, Transmit coil design for Wireless Power Transfer for medical implants, *2017 39th Annual International Conference of the IEEE Engineering in Medicine and Biology Society (EMBC)*, Seogwipo, 2017, pp. 2158–2161. https://doi.org/10.1109/EMBC.2017.8037282

16. A. Venkatasubramanian, B. Gifford, Modeling and design of antennas for implantable telemetry applications, *2016 38th Annual International Conference of the IEEE Engineering in Medicine and Biology Society (EMBC)*, Orlando, FL, 2016, pp. 6469–6472. https://doi.org/10.1109/EMBC.2016.7592210

17. D. Nikolayev, M. Zhadobov, P. Karban, R. Sauleau, Conformal antennas for miniature in-body devices: the quest to improve radiation performance. URSI Radio Sci. Bull. **2017**(363), 52–64 (2017). https://doi.org/10.23919/URSIRSB.2017.8409427

18. D. Nikolayev, Modeling and characterization of in-body antennas, *2018 IEEE 17th International Conference on Mathematical Methods in Electromagnetic Theory (MMET)*, Kiev, 2018, pp. 42–46. https://doi.org/10.1109/MMET.2018.8460279

19. M. Barbi, C. Garcia-Pardo, N. Cardona, A. Nevarez, V. Pons, M. Frasson, Impact of receivers location on the accuracy of capsule endoscope localization, *2018 IEEE 29th annual international symposium on personal, indoor and Mobile radio communications (PIMRC)*, Bologna, 2018, pp. 340–344. https://doi.org/10.1109/PIMRC.2018.8580862

20. C. Garcia-Pardo et al., Ultrawideband technology for medical in-body sensor networks: an overview of the human body as a propagation medium, phantoms, and approaches for propagation analysis. IEEE Antennas Propagat. Magaz. **60**(3), 19–33 (2018). https://doi.org/10.1109/MAP.2018.2818458

21. D. Nikolayev, M. Zhadobov, L. Le Coq, P. Karban, R. Sauleau, Robust Ultraminiature capsule antenna for ingestible and implantable applications. IEEE Trans. Antennas Propag. **65**(11), 6107–6119 (2017). https://doi.org/10.1109/TAP.2017.2755764

22. S. Perez-Simbor, C. Andreu, C. Garcia-Pardo, M. Frasson, N. Cardona, UWB path loss models for ingestible devices. IEEE Trans. Antennas Propagat.. https://doi.org/10.1109/TAP.2019.2891717

23. L. Chen et al., Radiofrequency propagation close to the human ear and accurate ear canal models, *40th Annual Int. Conf. of the IEEE Engineering in Medicine and Biology Society (EMBC 2018)*, Honolulu, HI, Jul. 17–21 2018

24. J.M. Elloian, G.M. Noetscher, S.N. Makarov, A. Pascual-Leone, Continuous wave simulations on the propagation of electromagnetic fields through the human head. IEEE Trans. Biomed. Eng. **61**, 1676–1683 (2014)

25. G.M. Noetscher et al., Comparison of cephalic and extracephalic montages for transcranial direct current stimulation – a numerical study. I.E.E.E. Trans. Biomed. Eng. **61**(9), 2488–2498 (2014)

26. S.N. Makarov, G.M. Noetscher, S. Arum, et al., Concept of a radiofrequency device for osteopenia/osteoporosis screening. Sci. Rep. **10**, 3540 (2020). https://doi.org/10.1038/s41598-020-60173-5

27. S. Perez-Simbor, C. Andreu, M. Frasson, N. Cardona, UWB path loss models for ingestible devices. IEEE Trans. Antennas Propagat. **67**(8), 5025–5034 (2019)

28. S.P. Simbor, *In-Body to on-Body Experimental UWB Channel Characterization for the Human Gastrointestinal Area*. PhD Thesis. Universitat Politecnica de Valencia. Spain. Oct. 2019

29. A. Prokop, T. Wittig, A. Morey, Using anatomical human body model for FEM SAR simulation of a 3T MRI system. Ch. 16. pp. 273–281, in *Brain and Human Body Modeling 2020*, vol. 2, (Springer Nature, New York, 2020) Open access. ISBN 978-3-030-45622

30. J.E. Brown, R. Qiang, P.J. Stadnik, L.J. Stotts, J.A. Von Arx, Calculation of MRI RF-induced voltages for implanted medical devices using computational human models. Ch. 16. pp. 283–294, in *Brain and Human Body Modeling 2019*, vol. 1, (Springer Nature, New York, 2019) Open access. ISBN 978-3-030-21293-3. Online: https://link.springer.com/chapter/10.1007/978-3-030-21293-3_14

31. J.E. Brown, R. Qiang, P.J. Stadnik, L.J. Stotts, J.A. Von Arx, RF-induced unintended stimulation for implantable medical devices in MRI. Ch. 17. pp. 283–292, in *Brain and Human Body Modeling 2020*, vol. 2, (Springer Nature, New York) Open access. ISBN 978-3-030-45622

32. H. Muranaka, T. Horiguchi, S. Usui, Y. Ueda, O. Nakamura, F. Ikeda, K. Iwakura, G. Nakaya, Evaluation of RF heating on humerus implant in phantoms during 1.5 T MRI imaging and comparisons with electromagnetic simulation. Magn. Reson. Med. Sci. **5**(2), 79–88 (2006)

33. H. Muranaka, T. Horiguchi, S. Usui, Y. Ueda, O. Nakamura, F. Ikeda, Dependence of RF heating on SAR and implant position in a 1.5T MR system. Magn. Reson. Med. Sci. **6**(4), 199–209 (2007)

34. H. Muranaka, T. Horiguchi, Y. Ueda, N. Tanki, Evaluation of RF heating due to various implants during MR procedures. Magn. Reson. Med. Sci. **10**(1), 11–19 (2011)

35. H. Homann, P. Börnert, H. Eggers, K. Nehrke, O. Dössel, I. Graesslin, Towards individualized SAR models and in vivo validation. Magn. Reson. Med. **66**(6), 1767–1776 (2011) PMID 21630346

36. H. Homann, SAR prediction and SAR management for parallel transmit MRI. Karlsruhe Translat. Biomed. Eng. **16**, 1–124 (2012)

37. M. Murbach, E. Neufeld, M. Capstick, W. Kainz, D.O. Brunner, T. Samaras, K.P. Pruessmann, N. Kuster, Thermal tissue damage model analyzed for different whole-body SAR and scan durations for standard MR body coils. Magn. Reson. Med. **71**(1), 421–431 (2014)

38. S. Oh, Y.-C. Ryu, G. Carluccio, C.T. Sica, C.M. Collins, Measurement of SAR-induced temperature increase in a phantom and in vivo with comparison to numerical simulation. Magn. Reson. Med. **71**(5), 1923–1931 (2014)

39. A. Lienhard, *Heat Transfer Textbook* (Phlogiston, Lexington, MA, 2005)

40. E.H. Wissler, Pennes' 1948 paper revisited. J. Appl. Phys. **85**, 35–41 (1998)

41. H. Arkin, Recent developments in modeling heat transfer in blood perfused tissues. I.E.E.E. Trans. Biomed. Eng. **41**(2), 97–107 (1994)

42. P.A. Hasgall, E. Neufeld, M.C. Gosselin, A. Klingenböck, N. Kuster, A. Klingenbock, P. Hasgall, M. Gosselin, IT'IS Database for thermal and electromagnetic parameters of biological tissues, Version 3.0, Sept. 1st, 2015. Available: www.itis.ethz.ch/database

43. Z. Xiaotong, L. Jiaen, B. He, Magnetic-resonance-based electrical properties tomography: a review. IEEE Rev. Biomed. Eng. **7**, 87–95 (2014)
44. E.K. Insko, L. Bolinger, Mapping of the radiofrequency field. J. Magn. Reson. A **103**(1), 82–85 (1993)
45. J. Wang, M. Qiu, Q.X. Yang, M.B. Smith, R.T. Constable, Measurement and correction of transmitter and receiver induced nonuniformities in vivo. Magn. Reson. Med. **53**(2), 408–417 (2005)
46. E. Cabot, A. Christ, N. Kuster, Whole body and local SAR in anatomical phantoms exposed to RF fields from birdcage coils. *Proceedings of the 29th General Assembly of the International Union of Radio Science*, 2008 August 7–16. Chicago, USA
47. S. Oh, A.G. Webb, T. Neuberger, B. Park, C.M. Collins, Experimental and numerical assessment of MRI-induced temperature change and SAR distributions in phantoms and in vivo. Magn Reson Med. **63**(1), 218–223 (2010) PMCID: PMC2836721
48. G. Bonmassar, P. Serano, L.M. Angelone, Specific absorption rate in a standard phantom containing a deep brain stimulation lead at 3 Tesla MRI, *2013 6th International IEEE/EMBS Conference on Neural Engineering (NER)*, San Diego, CA, 2013, pp. 747–750. https://doi.org/10.1109/NER.2013.6696042
49. C.M. Collins, S. Li, M.B. Smith, SAR and B1 field distribution in a heterogeneous human head model with a birdcage coil. Magn. Reson. Med. **40**(6), 847–856 (2005)
50. A. Rennings, L. Chen, S. Otto, D. Erni. B1-field inhomogeneity problem of MRI: basic investigation on a head- tissue-simulating cylinder phantom excited by a birdcage-mode. *42nd European Microwave Conference*, The Netherlands, Amsterdam, 2012 Nov 1; pp. 542–545
51. B.P. Tomas, H. Li, M.R. Anjum, Design and simulation of a birdcage coil using CST studio suite for application at 7T. IOP Conf. Ser. Mater. Sci. Eng. **51**(1), 1–6 (2013)
52. D.T.B. Yeo, Z. Wang, W. Loew, M.W. Volgel, I. Hancu, Local SAR in high pass birdcage and TEM body coils for multiple human body models in clinical landmark positions at 3T. J. Magn. Reson. Imaging **33**(5), 1209–1217 (2011) PMCID: PMC3081105
53. W. Liu, C.M. Collins, M.B. Smith, Calculation of B1 distribution, SAR, and SNR for a body-size birdcage coil loaded with different human subjects at 64 and 128 MHz. Appl. Magn. Reson. **29**(1), 5–18 (2005) PMID:23565039. PMCID: PMC3615460
54. L. Winter, F. Seifert, L. Zilberti, M. Murbach, B. Ittermann, MRI-related heating of implants and devices: a review. J. Magn. Reson. Imaging (2020). https://doi.org/10.1002/jmri.27194
55. M. Kozlov, R. Turner. RF transmit performance comparison for several MRI head arrays at 300 MHz. *Progress in Electromagnetic Research Symposium Proceedings.* 2013 March 28 Taipei, pp. 1052–1056
56. M. Kozlov, P.L. Bazin, H.E. Möller, N. Weiskopf. Influence of cerebrospinal fluid on specific absorption rate generated by 300 MHz MRI transmit array. *10th European Conference on Antennas and Propagation (EuCAP).* 2016 April, Davos, Switzerland, pp. 1–5

Part VI
High Frequency Electromagnetic Modeling: Microwave Imaging

Modeling and Experimental Results for Microwave Imaging of a Hip with Emphasis on the Femoral Neck

Johnathan Adams, Peter Serano, and Ara Nazarian

1 Introduction

Osteoporosis affects approximately 21.2% of women and 6.3% of men over the age of 50 world-wide [1]. In the United States alone, the estimated economic burden of osteoporosis-related fractures in 2005 was $17 billion and is expected to increase to about $25 billion by 2025 [7]. Hip fracture is one of the most serious and debilitating outcomes of osteoporosis [2, 3], with a 14–36% mortality rate during the first year following a fracture [4]. Hip fracture incidence rates increase exponentially with age in both women and men [5]. In 2010, there were estimated to be 158 million persons at high risk for a bone fracture, a staggering statistic expected to double by 2040 [6].

The World Health Organization (WHO) has defined individuals at risk for these fractures based on their areal Bone Mineral Density (aBMD, g/cm^2) relative to that of a normal young adult, as measured by Dual-energy X-ray Absorptiometry (DXA) [8]. However, DXA is not without its flaws. These include exposing patients to small ionizing radiation doses of up to 0.86 mrem [9]; measurement errors due to

J. Adams
Electrical & Computer Engineering Department, Worcester Polytechnic Institute, Worcester, MA, USA

P. Serano (✉)
Electrical & Computer Engineering Department, Worcester Polytechnic Institute, Worcester, MA, USA

Ansys, Inc., Canonsburg, PA, USA
e-mail: pete.serano@ansys.com

A. Nazarian
Musculoskeletal Translational Innovation Initiative, Carl J. Shapiro Department of Orthopaedic Surgery, Beth Israel Deaconess Medical Center, Harvard Medical School, Boston, MA, USA

Department of Orthopedic Surgery, Yerevan State Medical University, Yerevan, Armenia

© The Author(s) 2023
S. Makarov et al. (eds.), *Brain and Human Body Modelling 2021*,
https://doi.org/10.1007/978-3-031-15451-5_10

surrounding soft tissues [10, 11]; bone mineral density (BMD) measurements are affected by variations in bone size [12, 13]; and cortical and trabecular bone cannot be differentiated [14]. Additionally, fracture predictions based on aBMD are neither sensitive nor specific [15–19]. While DXA-based aBMD has been previously shown to be an important predictor of hip fracture risk [20], it does not offer a direct assessment of bone's load bearing capacity [21, 22]. Additionally, the predictive BMD value for fracture decreases in individuals over 70 years [23]. From the age of 60 to 80, the risk of hip fracture increases 13 times, while the decrease in BMD can only account for doubling of the risk [24]. Also, there is a wide overlap in BMD scores of postmenopausal women [25] who do and do not sustain osteoporotic fractures [26], and approximately 50% of fragility fractures occur in patients with DXA-derived BMD T-scores in the normal or low bone mass range [27, 28].

Quantitative ultrasound devices provide a low-cost, non-ionizing alternative to DXA using a specialized ultrasound transceiver to measure bones near the surface of the skin. A commercial example of this, Bindex®, uses the pulse-echo technique to measure the thickness of the frontal cortical shell of the tibia bone [29–32]. It has been found to correlate strongly with DXA measurements [29], a less than optimal gold standard.

1.1   Why Microwave Imaging

Microwave or radiofrequency imaging of (heel) bone was first introduced by Dr. Keith Paulsen and his research group at Dartmouth College in 2010, as an alternative non-ionizing diagnostic method to assess bone health [2, 33–36]. Due to the well-known complexity and poor spatial resolution of the standard microwave imaging setup [37, 38] used in these studies, they produced no clinically applicable results. However, the underlying physical idea of this method is simple and powerful. We have previously designed a simpler device to prove the viability of the concept at the wrist [39] and have achieved approximately 83% sensitivity and 94% specificity using a neural network classifier to differentiate between osteoporotic and healthy subjects [40].

1.2   Potential Difficulties

Taking these radiofrequency measurements is not without challenges. The transmission must pass through the bone in the region of interest and arrive at the receiver antenna with sufficient power to be measured, given a range of individuals. This makes measuring bones deeper in the body more difficult compared to more superficial bones. Additionally, the antennas must be placed such that the major component of the received signal is through the bone rather than its surrounding tissues.

1.3 Our Approach

This study consists of a set of simulations to determine field propagation inside the body validated by *in vivo* experimental measurements under the same conditions. The simulations produced models that included reflection coefficient S_{11} and transmission coefficient S_{21} in addition to the fields. These S-parameters can be measured in a physical setup using a network analyzer. The simulations and physical measurements were performed with the same antennas [41]. Additional simulations were performed with different antennas to investigate wideband measurements; these were not verified experimentally. The simulation results were analyzed primarily based on the electric field and Poynting vector.

2 Materials and Methods

This study was divided into two parts: first, a set of *in vivo* measurements using real antennas and second, a set of simulations using a corresponding human body model. The measurements were taken with Institutional Review Board (IRB) approval (IRB-19-0123) through Worcester Polytechnic Institute. The same human subject was used for all *in vivo* measurements.

2.1 Experimental Hardware

The antennas featured in this study are dual antiphase patch antennas [41] built using copper on FR4. Two sets of antennas, shown in Fig. 1, were investigated.

Set A (resonators: 2.0 cm × 1.4 cm, ground-plane: 5.0 cm × 1.9 cm) connected to matching networks that match them to 675 MHz. Matching networks were built with lumped components and applied at the antenna feeds, after the 180° power splitter (Mini-Circuits® ZFSCJ-2-232-S+, 5 MHz to 2.3 GHz).

Set B (resonators: 2.5 cm × 1.6 cm, ground-plane: 3.0 cm × 8.0 cm) were not matched to any particular frequency. The antenna feeds connected directly to the 180° power splitter (Mini-Circuits® ZFRSC-183-S+, DC to 1.8 GHz).

Both antennas had 0.5 cm spacing between the resonators. The antennas were connected to a Keysight FieldFox N9914A network analyzer. The network analyzer transmitted at −15 dBm over a frequency range of 30 kHz to 2 GHz at 401 points. The magnitude in dB and phase in degrees of S_{11} and S_{21} were saved to a CSV-file. The measurements were each a single frequency sweep.

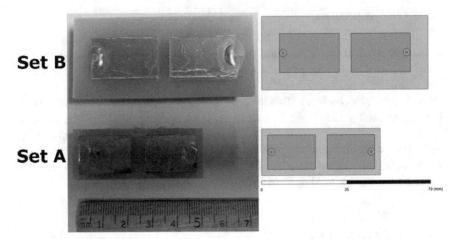

Fig. 1 Comparison of size between antennas from Set B (top) and from Set A (bottom). The left two are the physical antennas and the right two are the corresponding CAD models. The spacing between patches in both antennas is 0.5 cm. Antennas in Set A had resonators of 2 cm × 1.4 cm and a ground-plane of 5 cm × 1.9 cm. Antennas in Set B had resonators of 2.5 cm × 1.6 cm and a ground-plane of 3 cm × 8 cm. The antennas were fed from the back, the solder joints in the figure are the feeds

2.2   Measurement Sites

To test the viability of various sites for measuring transmission through the femoral neck, we first checked using both sets of antennas to determine if meaningful transmission could occur given the positions of the antennas. The exact positions investigated are shown in Fig. 2.

The positions investigated were:

1. On the side of the body, positioned over the greater trochanter.
2. On the side of the body, positioned next the iliac crest. The antenna in this position was rotated in the plane of the drawing in Fig. 2 to investigate different polarizations. The orientation shown in the figure (vertically aligned with the body and the antenna in position 1) was considered 0°, and rotation angles were measured toward the front of the body (clockwise on the right side, counterclockwise on the left side).
3. On the front of the body, positioned over the anterior superior iliac spine.
4. On the rear of the body, positioned over the top edge of the gluteus maximus.
5. On the front of the body, positioned horizontally in the same horizontal plane as the greater trochanter.
6. On the rear of the body, positioned horizontally and below the gluteus maximus.

Measurements were taken between two of the positions. The antennas were held to the body by the subject being measured, by pressing on the center of the ground plane of each antenna. This ensured deformation of the body so that the total length

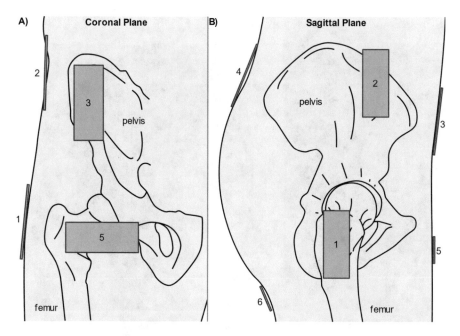

Fig. 2 (**a**) Front (coronal plane) view of the right side of the body with antenna positions by number. (**b**) Right side (sagittal plane) view of the body with antenna positions noted by number. In both, the skin profile, right pelvis, and right femur are shown in addition to the antennas. Antennas were pushed against the body such that gaps, such as the one near position 4, were not present during the measurement. The antenna in position 4 was located over the posterior iliac crest

of the antenna was contacting skin. The positions that were not in use for a given measurement did not have antennas present. All position combinations measured were measured with both Set A and Set B antennas.

2.3   Simulated Antenna Positioning and Human Body Model

Antenna positions on the simulated body model were the same as those on the *in vivo* model and are shown in Fig. 3. The base CAD model is the Ansys male human model. It was chosen to match the *in vivo* subject, who is male.

The CAD model includes the full body with bones, muscles, and fat modelled throughout. Some skin layers and fat deposits are represented in aggregate by a volume with the average electric properties of the human body [42, 43]. Cartilage in joints, such as the hip joint, is not modelled by default. We investigated the effects of cartilage by producing a new shell using the area between the femur and pelvis making up the ball joint. This new volume was between 3 to 10 mm thick due mostly to the large-triangle tessellation of the bones' shells. In addition, two outer skin shells were added with properties derived from the VHP-Female v.5.0 model [44].

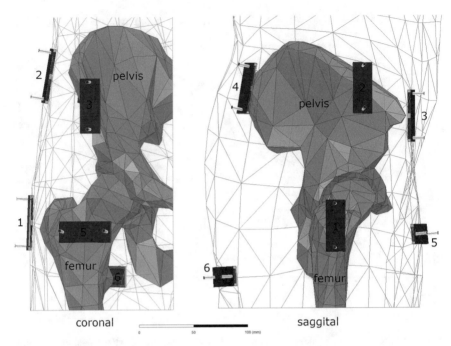

Fig. 3 (**a**) Front (coronal plane) view of the right side of the CAD model with antenna positions by number. (**b**) Right side (sagittal plane) view of the CAD model with antenna positions noted by number. In both, the wireframe body shell, right pelvis, and right femur are shown in addition to the antennas. The body shell was flattened or Boolean-subtracted using the antenna's shape to eliminate gaps and ensure good coupling at each position. The apparent difference in location of position 4 is due to perspective of the drawing in Fig. 2

2.4 Software Modelling of Matched Antennas

The matching networks were modelled in Ansys using S-Parameter measurements of the physical matching networks. The matching networks' measurements were taken over the same frequency range (30 kHz to 2 GHz) and with the same resolution (401 points) as the *in vivo* transmission measurements. However, they were taken at 0 dBm and averaged across 8 sweeps, whereas the *in vivo* measurements were taken at a lower power (−15 dBm, see above) and with only a single frequency sweep. The input ports were all 50 Ω characteristic impedance, identical to the physical network analyzer. The 180° power splitters were modelled with an ideal splitter model. Figure 4 shows a typical simulation configuration for a single antenna at position 1 compared to a physical measurement at the same site.

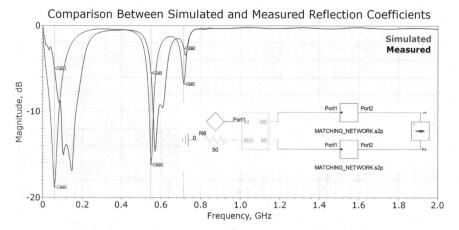

Fig. 4 Comparison of reflection coefficient magnitude |S_{11}| for the simulated antenna (red) and the in-vivo antenna (dashed black). This figure additionally shows the configuration of the matching and power splitting circuits in HFSS for the simulated curve. Both the simulated and measured curves were produced from antennas at position 1

3 Results

This section is divided between the *in vivo* and simulated results. The *in vivo* results provide an assessment of the total transmission in a real subject, given a set of antenna positions, and the simulations show where inside the body the transmission occurs.

3.1 In Vivo *Measurements*

The positions shown in Fig. 2 are positions between which transmission was achieved. Additional sites were measured, including one between sites 4 and 6 through the center of the gluteus maximus, but no meaningful signal was received. Figure 5 shows the transmission coefficient for a selection of antenna position pairs using Set A, while Fig. 6 shows the transmission coefficient for the same position pairs using Set B.

In addition to the measurements shown in the figures, transmission from position 1 to position 2 was measured with varying polarizations achieved by rotating the position 2 antenna in 45° increments. For Set A, the highest average transmission over the bandwidth of the antenna transmission was seen at 45° of rotation while the lowest at 135°. Set B was less consistent but showed similar results: minimum transmission at 270° and maximum at either 45° (on the left side of the body) or 135° (on the right side). Overall, Set B showed less change in S_{21} over various angles of rotation than Set A, potentially due to noise. Set A showed differences of

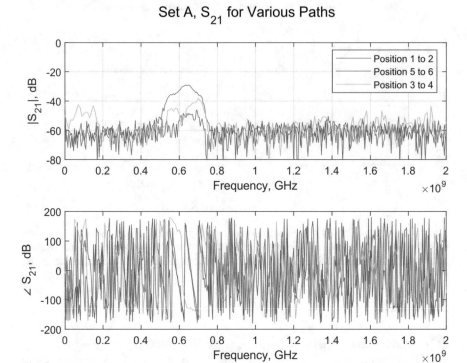

Fig. 5 Comparison of transmission coefficient S_{21} when using antennas from Set A for three propagation paths: First, position 1 to position 2, a semicircular path through the compartment. Second, position 5 to position 6, a straight path through the upper femur. Third, position 3 to position 4, a straight path through the upper pelvis

S_{21} at the same frequency within the passband of the antenna with different polarizations up to 20 dB, while Set B showed differences up to only 10 dB. Lying down during the measurement process decreased this variation by about half. Table 1 shows the maximum measured magnitude of the transmission coefficient for each orientation tested, for both sets of antennas.

3.2 Simulations

First, the relative agreement between the simulated and measured results is characterized by Fig. 7, in which there are resonances at approximately the same frequencies in the measured and simulated environments, but the simulated environment experiences significantly more attenuation on transmission than the measured environment.

Next, the propagation paths of the waves were observed using animated electric field plots in various observation planes and 3-D Poynting vector plots in the bones

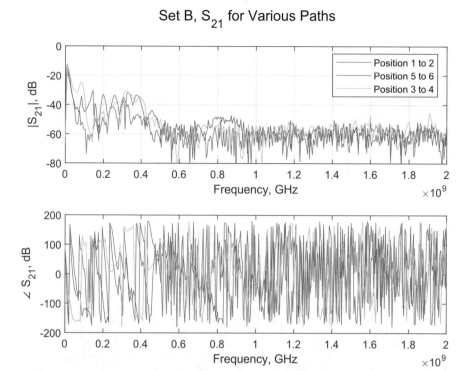

Fig. 6 Comparison of transmission coefficient S_{21} when using antennas from Set B for three propagation paths: First, position 1 to position 2, a semicircular path through the compartment. Second, position 5 to position 6, a straight path through the upper femur. Third, position 3 to position 4, a straight path through the upper pelvis

Table 1 Maximum transmission coefficient magnitude $|S_{21}|$ for various relative polarizations using both sets of antennas on the right leg. Measured from position 1 to position 2

Angle, degrees	Set A, right leg		Set B, right leg	
	f, MHz	$|S_{21}|$, dB	f, MHz	$|S_{21}|$, dB
0 (co-polarized)	615.0	−38.663	10.0	−12.254
45	635.0	−29.127	10.0	−12.066
90 (cross-polarized)	695.0	−42.658	10.0	−13.860
135	705.0	−39.581	10.0	−11.955
180 (co-polarized)	705.0	−37.817	10.0	−12.741
225	600.0	−36.861	10.0	−12.891
270 (cross-polarized)	710.0	−39.151	10.0	−14.000
315	705.0	−37.694	10.0	−12.057

and the body. At higher frequencies, a surface-propagating wave is present, as seen in Fig. 8. The Poynting vector plots in the femur and pelvis for the three transmission configurations in Figs. 5, 6, and 7 are shown in Fig. 9.

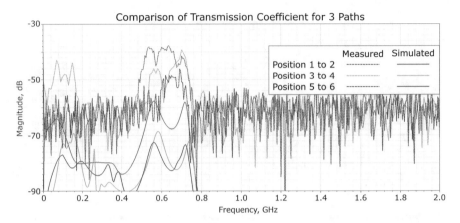

Fig. 7 Comparison of transmission coefficient S_{21} when using antennas from Set A for three propagation paths: First (red): position 1 to position 2, a semicircular path through the compartment. Second (blue): position 5 to position 6, a straight path through the upper femur. Third (green): position 3 to position 4, a straight path through the upper pelvis. The dashed lines are the measured in-vivo $|S_{21}|$ (also seen in Fig. 5) and the solid lines are simulated

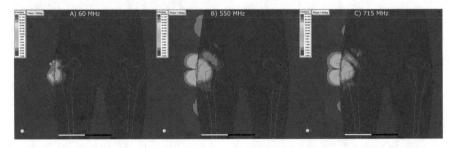

Fig. 8 Electric field magnitude in the sagittal plane at different frequency bands. (**a**) is 60 MHz, (**b**) is 550 MHz, and C is 715 MHz. All three are snapshots from animations, taken at a phase of 60°. Note the vertical surface-propagating wave is present in **b** and **c** but not in **a**. The antennas for these measurements are the Set A antennas, located at position 1

In addition to the results shown in the figures, simulations were performed with a dielectric "belt" between the transmitting and receiving antennas to attenuate the surface wave. The effect was not strong enough to reduce the magnitude of the surface wave to a level comparable to that of the wave propagating through the bone.

4   Discussion

4.1   Limitations

This study only considered the strongest component of the received wave. Simulations suggest that this component likely propagates through skin, fat, and muscle when the antennas are on the same side of the body compartment. The same

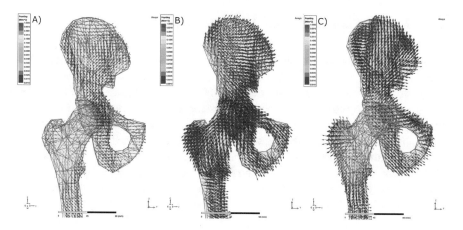

Fig. 9 Poynting vector distribution in the femur and pelvis for three antenna position pairs: (**a**) transmission from position 1 to position 2, (**b**) transmission from position 3 to position 4, and (**c**) transmission from position 4 to position 5. Poynting vector magnitude is represented by color, warmer is larger

simulations also suggest propagation occurs through the bone and this second component will accrue some phase shift (delay) relative to the one that propagates through the soft tissue.

This study performed *in vivo* measurements on only one subject, a 26-year-old male, who is not at significant risk of osteoporosis according to standard risk factors.

In vivo spectra were determined from a single frequency sweep; therefore, noise contents in the spectra are more significant than had the measurements been performed using averaging of multiple sweeps.

Matched (Set A) and unmatched (Set B) antennas are not identical and have different resonant frequencies. Set A had a bandwidth of about 230 MHz, centered on about 675 MHz (when matched) and Set B had a bandwidth of about 420 MHz, centered on about 215 MHz.

4.2 Validation of Simulation Using In Vivo Results

While the measured and simulated results are not a perfect match, the resonant frequencies are collocated in the two spectra for the same antenna set and positions. Some of the difference in transmission coefficient magnitude between the measured and simulated spectra is due to differences between the model and the physical subject. These differences include the level of detail of the CAD model, and differences in physical shape between the CAD model and the subject.

4.3   Propagation Paths and Antenna Position

To achieve transmission through the bone, antennas should be placed on the opposite sides of the body compartment. If placed on the same side of the body compartment, the surface wave has the shortest path between the two antennas and thereby dominates the received signal. Direct transmission across a body compartment, contrarily, puts the shortest path between the antennas through the bone at the center of the compartment, and the Poynting vector for such a setup is the largest at the center of the compartment [39]. This is illustrated in Fig. 9, where the distribution in part C shows more even transmission through the femoral neck than part A. Part C's antennas are transmitting across the compartment while part A's transmit in a u-shape, starting and ending on the same side of the compartment.

4.4   Frequency Choice for Propagation Through Bone

It is common knowledge that lower frequencies provide better human body penetration but lower spatial resolution in microwave imaging. Our simulations have confirmed this, but we also note that in this application we can consider frequencies that are lower than traditionally considered for microwave imaging of the human body, due to the independence of this approach from spatial features. Therefore, a frequency of operation closer to 60 MHz with a reduced surface wave is preferable in this application to a higher frequency. Any waveguide-like effects from higher-frequency waves propagating through bones are overshadowed by the lack of penetration to reach the bones in the first place, and by the large surface-propagating waves produced by these high frequencies.

References

1. J.A. Kanis, E.V. McCloskey, H. Johansson, A. Oden, L.J. Melton, N. Khaltaev, A reference standard for the description of osteoporosis. Bone **42**(3), 467–475 (2008)
2. P.M. Meaney et al., Clinical microwave tomographic imaging of the calcaneus: a first-in-human case study of two subjects. IEEE Trans. Biomed. Eng. **59**(12), 3304–3313 (Dec. 2012). https://doi.org/10.1109/TBME.2012.2209202
3. S. El-Kaissi et al., Femoral neck geometry and hip fracture risk: The Geelong osteoporosis study. Osteoporos. Int. **16**(10), 1299–1303 (2005). https://doi.org/10.1007/s00198-005-1988-z
4. S. Mundi, B. Pindiprolu, N. Simunovic, M. Bhandari, Similar mortality rates in hip fracture patients over the past 31 years. Acta Orthop. **85**(1), 54–59 (2014). https://doi.org/10.3109/17453674.2013.878831
5. P. Kannus, A. Natri, T. Paakkala, M. Järvinen, An outcome study of chronic patellofemoral pain syndrome. Seven-year follow-up of patients in a randomized, controlled trial. J. Bone Joint Surg. **81**(3), 355–363 (1999). https://doi.org/10.2106/00004623-199903000-00007
6. A. Odén, E.V. McCloskey, J.A. Kanis, N.C. Harvey, H. Johansson, Burden of high fracture probability worldwide: secular increases 2010–2040. Osteoporos. Int. **26**(9), 2243–2248 (2015)

7. R. Burge, B. Dawson-Hughes, D.H. Solomon, J.B. Wong, A. King, A. Tosteson, Incidence and economic burden of osteoporosis-related fractures in the United States, 2005–2025. J. Bone Miner. Res. **22**(3), 465–475 (2006)

8. J.A. Kanis, *Assessment of Osteoporosis at the Primary Health Care Level* (University of Sheffield, Sheffield, UK, rep., 2007)

9. S. Lee, D. Gallagher, Assessment methods in human body composition. Curr. Opin. Clin. Nutr. Metab. Care **11**(5), 566–572 (2008). https://doi.org/10.1097/mco.0b013e32830b5f23

10. E. Lochmüller, N. Krefting, D. Bürklein, F. Eckstein, Effect of fixation, soft-tissues, and scan projection on bone mineral measurements with dual energy X-ray absorptiometry (DXA). Calcif. Tissue Int. **68**(3), 140–145 (2001). https://doi.org/10.1007/s002230001192

11. O. Svendsen, C. Hassager, V. Skødt, C. Christiansen, Impact of soft tissue on in vivo accuracy of bone mineral measurements in the spine, hip, and forearm: a human cadaver study. J. Bone Miner. Res. **10**(6), 868–873 (2009). https://doi.org/10.1002/jbmr.5650100607

12. E. Lochmüller, P. Miller, D. Bürklein, U. Wehr, W. Rambeck, F. Eckstein, In situ femoral dual-energy X-ray absorptiometry related to ash weight, bone size and density, and its relationship with mechanical failure loads of the proximal femur. Osteoporos. Int. **11**(4), 361–367 (2000). https://doi.org/10.1007/s001980070126

13. A. Prentice, T. Parsons, T. Cole, Uncritical use of bone mineral density in absorptiometry may lead to size-related artifacts in the identification of bone mineral determinants. Am. J. Clin. Nutr. **60**(6), 837–842 (1994). https://doi.org/10.1093/ajcn/60.6.837

14. H. Genant et al., Noninvasive assessment of bone mineral and structure: state of the art. J. Bone Miner. Res. **11**(6), 707–730 (2009). https://doi.org/10.1002/jbmr.5650110602

15. S. Schuit et al., Fracture incidence and association with bone mineral density in elderly men and women: the Rotterdam study. Bone **34**(1), 195–202 (2004). https://doi.org/10.1016/j.bone.2003.10.001

16. S. Cummings et al., Improvement in spine bone density and reduction in risk of vertebral fractures during treatment with antiresorptive drugs. Am. J. Med. **112**(4), 281–289 (2002). https://doi.org/10.1016/s0002-9343(01)01124-x

17. R. Heaney, Is the paradigm shifting? Bone **33**(4), 457–465 (2003). https://doi.org/10.1016/s8756-3282(03)00236-9

18. D. Bauer et al., Change in bone turnover and hip, non-spine, and vertebral fracture in alendronate-treated women: the fracture intervention trial. J. Bone Miner. Res. **19**(8), 1250–1258 (2004). https://doi.org/10.1359/jbmr.040512

19. B. Riggs, L. Melton, Bone turnover matters: the Raloxifene treatment paradox of dramatic decreases in vertebral fractures without commensurate increases in bone density. J. Bone Miner. Res. **17**(1), 11–14 (2002). https://doi.org/10.1359/jbmr.2002.17.1.11

20. C. Gomez Alonso, M. Diaz Curiel, F. Hawkins Carranza, R. Perez Cano, A. Diez Perez, Femoral bone mineral density, neck-shaft angle and mean femoral neck width as predictors of hip fracture in men and women. (in eng), Osteoporos. Int. **11**(8), 714–720 (2000)

21. T. Hoc, L. Henry, M. Verdier, D. Aubry, L. Sedel, A. Meunier, Effect of microstructure on the mechanical properties of Haversian cortical bone. (in eng), Bone **38**(4), 466–474 (2006)

22. M. Viswanathan et al., Screening to prevent osteoporotic fractures: updated evidence report and systematic review for the US preventive services task force. JAMA **319**(24), 2532–2551 (2019)

23. H. Johansson, J.A. Kanis, A. Oden, O. Johnell, E. McCloskey, BMD, clinical risk factors and their combination for hip fracture prevention. (in eng), Osteoporos. Int. **20**(10), 1675–1682 (2009)

24. J.A. Kanis et al., A meta-analysis of previous fracture and subsequent fracture risk. (in eng), Bone **35**(2), 375–382 (Aug 2004)

25. S.A. Wainwright et al., Hip fracture in women without osteoporosis. (in eng), J. Clin. Endocrinol. Metab. **90**(5), 2787–2793 (May 2005)

26. R.E. Small, Uses and limitations of bone mineral density measurements in the management of osteoporosis. (in eng), Med. Gen. Med. **7**(2), 3 (2005)

27. E.S. Siris et al., The effect of age and bone mineral density on the absolute, excess, and relative risk of fracture in postmenopausal women aged 50–99: results from the National Osteoporosis Risk Assessment (NORA). Osteoporos. Int. **17**(4), 565–574 (2006)
28. P. Choksi, K.J. Jepsen, G.A. Clines, The challenges of diagnosing osteoporosis and the limitations of currently available tools. Clin. Diabetes Endocrinol. **4** (2018)
29. J. Karjalainen, O. Riekkinen, H. Kröger, Pulse-echo ultrasound method for detection of postmenopausal women with osteoporotic BMD. Osteoporos. Int. **29**(5), 1193–1199 (2018). https://doi.org/10.1007/s00198-018-4408-x
30. J. Karjalainen et al., Multi-site bone ultrasound measurements in elderly women with and without previous hip fractures. Osteoporos. Int. **23**(4), 1287–1295 (2011). https://doi.org/10.1007/s00198-011-1682-2
31. J. Karjalainen, O. Riekkinen, J. Töyräs, J. Jurvelin, H. Kröger, New method for point-of-care osteoporosis screening and diagnostics. Osteoporos. Int. **27**(3), 971–977 (2015). https://doi.org/10.1007/s00198-015-3387-4
32. J. Karjalainen, O. Riekkinen, J. Toyras, H. Kroger, J. Jurvelin, Ultrasonic assessment of cortical bone thickness in vitro and in vivo. IEEE Trans. Ultrason. Ferroelectr. Freq. Control **55**(10), 2191–2197 (2008). https://doi.org/10.1109/tuffc.918
33. T. Zhou, P.M. Meaney, M.J. Pallone, S. Geimer, K.D. Paulsen, Microwave tomographic imaging for osteoporosis screening: a pilot clinical study, *2010 Annual International Conference of the IEEE Engineering in Medicine and Biology*, Buenos Aires, Argentina, 2010, pp. 1218–1221. https://doi.org/10.1109/IEMBS.2010.5626442
34. P.M. Meaney, D. Goodwin, A. Golnabi, M. Pallone, S. Geimer, K.D. Paulsen, 3D microwave bone imaging, *2012 6th European Conference on Antennas and Propagation (EUCAP)*, Prague, Czech Republic, 2012, pp. 1770–1771. https://doi.org/10.1109/EuCAP.2012.6206024
35. A.H. Golnabi, P.M. Meaney, S. Geimer, T. Zhou, K.D. Paulsen, Microwave tomography for bone imaging, *2011 IEEE International Symposium on Biomedical Imaging: From Nano to Macro*, Chicago, IL, USA, 2011, pp. 956–959. https://doi.org/10.1109/ISBI.2011.5872561
36. P. Meaney, T. Zhou, D. Goodwin, A. Golnabi, E. Attardo, K. Paulsen, Bone dielectric property variation as a function of mineralization at microwave frequencies. Int. J. Biomed. Imaging **2012**, 1–9 (2012). https://doi.org/10.1155/2012/649612
37. R. Chandra, H. Zhou, I. Balasingham, R.M. Narayanan, On the opportunities and challenges in microwave medical sensing and imaging. IEEE Trans. Biomed. Eng. **62**(7), 1667–1682 (2015). https://doi.org/10.1109/TBME.2015.2432137
38. V. Zhurbenko, T. Rubæk, V. Krozer, P. Meincke, Design and realisation of a microwave three-dimensional imaging system with application to breast-cancer detection. IET Microw. Antennas Propagat. **4**(12), 2200 (2010). https://doi.org/10.1049/iet-map.2010.0106
39. S. Makarov, G. Noetscher, S. Arum, R. Rabiner, A. Nazarian, Concept of a radiofrequency device for osteopenia/osteoporosis screening. Sci. Rep. **10**(1) (2020). https://doi.org/10.1038/s41598-020-60173-5
40. J.W. Adams, Z. Zhang, G.M. Noetscher, A. Nazarian, S.N. Makarov, Application of a neural network classifier to radiofrequency-based osteopenia/osteoporosis screening. IEEE J. Translat. Eng. Health Med. **9**, 1–7 (2021., Art no. 4900907). https://doi.org/10.1109/JTEHM.2021.3108575
41. S. Makarov, A. Nazarian, W. Appleyard, G. Noetscher, Microwave antenna array and testbed for osteoporosis detection, US Patent # 10,657,338, May 19, 2020. Available: https://patents.justia.com/patent/10657338
42. C. Gabriel, *Compilation of the Dielectric Properties of Body Tissues at RF and Microwave Frequencies*, Air Force Materiel Command, Brooks Air Force Base, 1996

43. P.A. Hasgall, F. Di Gennaro, C. Baumgartner, E. Neufeld, B. Lloyd, M.C. Gosselin, D. Payne, A. Klingenböck, N. Kuster, IT'IS Database for thermal and electromagnetic parameters of biological tissues, Version 4.0, May 15, 2018. https://doi.org/10.13099/VIP21000-04-0. Available: itis.swiss/database
44. G.M. Noetscher, P. Serano, W.A. Wartman, K. Fujimoto, S.N. Makarov, Visible human project® female surface based computational phantom (Nelly) for radio-frequency safety evaluation in MRI coils. *PLOS One* **16**(12) (2021). https://doi.org/10.1371/journal.pone.0260922

Index

A

Active implantable medical devices (AIMDS),
125, 126, 129

B

Biological tissue heating, 138
Biomedical signal analysis and
propagation, 136
Birdcage MRI coil, 143, 144
Bone mass density, 136
Boundary element fast multipole method,
53, 62, 76
Boundary Element Method accelerated by the
Fast Multipole Method (BEM-FMM),
62, 64, 66, 67, 76, 80

C

Cell membrane polarization, 61
Computational human model (CHM),
125–130, 135, 136, 141, 144, 146, 147
Computational modeling, 75, 82, 85
Computational model validation, 75
Concomitant fields, 39, 52–53, 56
Craniotomy, 21

E

Eddy currents, 39, 45, 47, 51, 52, 56
Electrical conductivities, 40, 53, 64,
90, 102–121
Electric field intensities, 19, 20, 138

Electric field magnitude and orientation, 82,
83, 85, 138, 164
Electrode montages, 102, 104, 106, 110
Experimental validation, 93–95

F

Finite element analysis, 90, 91
Finite element method (FEM), 20, 22, 25, 40,
53, 62, 105, 133–148
Freesurfer, 69, 76, 77, 80

G

Glioblastoma (GBM), 3–16, 19, 20, 33, 34

H

headreco, 76–82, 84, 85
Human body modeling, 39–56, 159–160
Human brain stimulation, 90, 96
Human cranial modeling, 140

I

Image segmentation methods, 75–86
Induced electric fields, 4, 25, 61, 91, 93,
95, 96
In vivo experimental validation, 157

J

Joule heating, 14

M
Magnetic stimulation profile (MSP), 62,
      63, 66–68
Medical device development tool (MDDT),
      134, 139–148
Microwave imaging, 155–166
Model mesh co-registration, 139
MRI gradient coils, 40
mri2mesh, 76–85
MR safety assessment, 130
Multichannel transcranial magnetic
      stimulation (MTMS), 61–71
Multi-layer winding, 91, 92

N
Nerve trajectories, 43–46, 48, 51, 54–56
Neurodynamic simulations, 39, 43–45
Neuromodulation, 86, 102, 106, 119, 120
Non-invasive brain stimulation, 75
Non-invasive stimulation, 101, 120
Novotal, 4, 7–9, 16

O
Optune, 3, 5, 6, 8, 10
Osteoporosis, 134, 155, 165

P
Patch antenna array, 157
Peripheral nerve stimulation (PNS), 39–56

R
Realistic human body models, 102, 103
Real-time neuronavigation, 61–71
Recorded motor evoked potentials
      (MEPs), 69, 70
RF-induced heating, 125–130, 134

S
Sensorimotor functions, 120
Skullremodeling surgery (SR-surgery), 19–34
Specific absorption rate (SAR), 53, 127, 128,
      134, 135, 137–138, 140–143, 148
Stimulation depth-focality tradeoff, 90–92, 96
Suprathreshold stimulation, 69, 70
Synaptic communication, 120

T
Temperatures, 3–16, 126–128, 133–148
Tissue simulating media (TSM), 127–130
TMS coil array, 63
TMS coil designs, 90, 96
Transcranial magnetic stimulation (TMS),
      61–67, 69, 70, 75–86, 89, 90, 92,
      94–96, 118
Transcutaneous spinal direct current
      stimulation (tsDCS), 101–121
Transfer function method, 125
Transmembrane potential, 43, 44, 102
Tumor, 3–16, 19–34, 118
Tumor treating fields (TTfields), 3–16, 19–34

Printed in the United States
by Baker & Taylor Publisher Services